Selected Titles in This Series

W0016848

(Continued in the back of this publication)

Hyperbolic Partial Differential Equations and Wave Phenomena

Translations of

MATHEMATICAL MONOGRAPHS

Volume 189

Hyperbolic Partial Differential Equations and Wave Phenomena

Mitsuru Ikawa

Translated by
Bohdan I. Kurpita

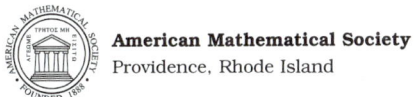

American Mathematical Society
Providence, Rhode Island

IWANAMI SERIES IN MODERN MATHEMATICS

偏微分方程式 2

PARTIAL DIFFERENTIAL EQUATIONS 2

by Mitsuru Ikawa

Copyright © 1997 by Mitsuru Ikawa

Originally published in Japanese
by Iwanami Shoten, Publishers, Tokyo, 1997

Translated from the Japanese by Bohdan I. Kurpita

1991 *Mathematics Subject Classification.* Primary 35L10;
Secondary 35L20, 35P25.

Library of Congress Cataloging-in-Publication Data

Ikawa, Mitsuru, 1942–
  [Henbibun hoteishiki 2. English]
  Hyperbolic partial differential equations and wave phenomena / Mitsuru Ikawa ;
translated by Bohdan I. Kurpita.
     p. cm. — (Translations of mathematical monographs ; v. 189)(Iwanami
series in modern mathematics)
  Includes bibliographical references and index.
  ISBN 0-8218-1021-9 (alk. paper)
  1. Differential equations, Hyperbolic. 2. Boundary value problems. 3. Wave
equation. I. Title. II. Series. III. Series: Iwanami series in modern mathematics
QA377.I4713   2000
515′.353—dc21   00-025700
                                                                      CIP

10 9 8 7 6 5 4 3 2 1   05 04 03 02 01 00

Contents

Preface to the English Edition

This book is a translation into English of a book that I wrote in Japanese. This book is based on my lectures at various universities.

When I originally set pen to paper, I had only a Japanese audience in mind. So, I am deeply pleased and honored that by means of this English translation the material in this book will reach a wider audience.

I would like to express my appreciation to the American Mathematical Society for publishing this translation. My heartfelt thanks go to Bohdan I. Kurpita for his translation of my book from Japanese into English.

August 1999

Mitusuru Ikawa

Preface to the Japanese Edition

In this book we discuss initial boundary value problems for second order hyperbolic equations. The term hyperbolic equation refers to members of a specific class of partial differential equations. The most representative examples of this class are the partial differential equations that describe wave phenomena. In essence, the study of hyperbolic equations and the mathematical investigation of wave phenomena can be thought of as one and the same thing. For the purposes of this book, we will restrict our attention to linear equations. What this means for wave phenomena is that the oscillations cannot be particularly large.

Most likely, the first image the mind conjures up when one hears or reads the word "wave" is of water lapping back and forth. More precisely, the surface of the water undulates in a periodic manner with the effect being passed on to the surroundings. This type of phenomenon can also be seen when a string vibrates or a membrane oscillates.

Another form of "wave" behaviour that readily comes to mind is that of sound propagating through air or some other substance. Another example is an electromagnetic wave. A less sanguine example is that of an earthquake, which is a wave that radiates outwards from an epicenter and, on occasion, the subsequent oscillations have a devastating effect causing widespread destruction and bringing misery and death.

As the above examples illustrate, we encounter many wave phenomena on a daily basis, some wave phenomena can be readily seen and some not. At first sight, these wave phenomena seem disparate since they arise from different physical circumstances, however, if we write the mathematical formulae that govern these phenomena, then the phenomena all conform to partial differential equations of the same form; namely, *hyperbolic equations*.

The added benefit of expressing these types of phenomena mathematically is that investigating the partial differential equations leads to detailed insight into wave phenomena. In addition, the similarity among the partial differential equations for various wave phenomena suggests a deep commonality between the phenomena.

In the case of the equation for the propagation of sound, the unknown function is a scalar function. On the other hand, in Maxwell's equations, which govern the transmission of electromagnetic waves, the unknown function is vector valued. This implies that even though the hyperbolic equations are of the same type, the complexity of the associated phenomena will differ according to whether the unknown function is scalar valued or vector valued. Thus, by focusing our deliberations on the characteristics of each of the equations, we will unearth properties of both sound and electromagnetism.

Throughout this book, linear hyperbolic equations of second order with a scalar-valued unknown function will be central to our discussions. The reason is that this type of equation is common and fundamental for all of the equations that describe wave phenomena. With this in mind, our overarching aim is to consider particular linear hyperbolic equations of second order and elucidate the properties of the phenomena governed by this second order equation. In this way, we will discover properties that are common to various wave phenomena.

In summary, we will concern ourselves with a wave transmitted in a space with boundary, and, having been given the initial condition(s) and the state on the boundary, we will investigate how the solution develops with time. Towards this end, we first need to give a mathematical proof of the existence of a solution for the given problem. Then, our core problems are to clarify mathematically observations such as the direction of motion, the reflection of the wave at a boundary, the refraction of the wave at the surface of two substances that are in contact, etc. The clarification of these observations will be obtained by determining properties of the solution.

In closing, I wish to express my deep gratitude to the various members of the editorial board who encouraged me to write this book. In particular, I would like to extend my heartfelt appreciation to Professor Aomoto who read the original manuscript and offered many helpful suggestions.

Also, I wish to thank the editorial staff of the publisher, Iwanami. I have great admiration for all their efforts.

January 1997

Mitusuru Ikawa

Outline of the Theory and Objectives

As explained in the preface, the primary objective of this book is to understand wave phenomena in mathematical terms.

So immediately in Chapter 1, we develop the mathematical laws that govern the behaviour for several wave phenomena. To be more precise, the mathematical expressions that we derive in this first chapter consist of a relationship or relationships between the partial derivatives of the unknown function(s); such a relationship(s) is usually referred to as a *partial differential equation (or system of equations)*. Among the phenomena to be considered in Chapter 1 are the vibration of a string, the oscillation of a spring and the propagation of sound.

As illustrative examples, we look at the vibration of a string and the oscillation of a spring. In the case of a string, if the string is disturbed from its state of rest, then the tension of the string causes a force to arise that pulls the disturbed part towards its rest position. In fact, this force gives a certain acceleration to the string, and the vibration is passed on to the surroundings. Thus a wave is formed.

While for the case of a spring, instead of tension we need to investigate the effect of elasticity. Briefly, what we mean is that when we apply a force to some part of the spring, we effect a contraction on this part of the spring, which, in turn, passes an oscillation to the surrounding parts.

Even the casual observer should be able to perceive a certain dichotomy between the nature of the vibration and the oscillation described above. In the case of the vibration of a string, each part of the string vibrates perpendicular to the direction in which the wave moves. In other words, each part of the string vibrates in a direction that is perpendicular to its position at rest, and with time the vibration is transmitted along the stretched string in the form of a wave. Hence, this kind of wave is known as a *transverse wave*.

In contrast, each part of a spring oscillates in the direction of the extended spring. So, this kind of wave is called a *longitudinal wave*.

From the discussion above, it is clear that the respective waves arise from quite different physical conditions, and the vibration and oscillation, themselves, are substantially different. However, as we will see in Chapter 1, if we derive the respective partial differential equations for these two phenomena, what we see is that the differential equations are exactly the same. This means that even though the vibration/oscillation are perceived differently, because these two waves have the same equation, a property for one of the waves also becomes a property of the other wave. For example, the speed of propagation to its surroundings is constant for each of them, reflection at its end point occurs for each of them, etc.

Continuing on from the introduction of the equation of the wave for the string and spring, we next turn our attention to the equation for the propagation of sound. Even if the dimension of the space is high, the astute reader will not be surprised to learn that the partial differential equation which will be found has the same form as that for a string or a spring. Therefore, an immediate consequence is that the propagation of sound must share the same properties, two of which were just mentioned, as the vibration of a string and the oscillation of a spring.

Having established the above, we will next write down, but not derive, *Maxwell's equations* that describe the propagation of electromagnetic waves and *elastic waves* that describe how a wave is transmitted through an elastic body, one example of such a wave comes from the observation of earthquakes. In both of these cases, the equations have unknown functions that are vector valued. Therefore, the oscillating phenomena that are governed by these equations include rather complicated terms. However, by careful consideration of these equations, it is possible to see that they consist of several equations each of which gives an expression as for the propagation of sound, and so without equivocation we can call them *wave equations*. In fact, if we restrict ourselves to relatively simple conditions, then we are able to see that each part of the unknown function becomes a solution of the wave equation. Conversely, by using the solution of the wave equation, we can construct solutions of the above equations. In the above sense, as stated in the Preface, the study of a single hyperbolic equation forms a basis for the investigation of a wave.

Also in Chapter 1, we define what we mean by a linear hyperbolic equation of second order. Then, in Chapter 2, we discuss the basic properties of the solution of this equation. As one of the more prominent properties of the solution, we will find that the speed of propagation of the solution is finite. For the wave equation we know that the solution propagates with a fixed speed, and thus we can conclude that a common property of hyperbolic equations is that the speed with which the oscillation/vibration, given at some part, propagates to its surroundings is finite.

Having completed the above, we will prove that the solution for the initial boundary value problem exists and is smooth.

In Chapter 3, we consider the asymptotic solution of the initial boundary value problem for hyperbolic equations. As stated above, in Chapter 2 we prove the existence of the solution. The proof depends on methods from functional analysis. However, when the state is fixed, these methods are not very suitable for studying the detailed behaviour of the solution. On the other hand, using the methods of asymptotic solutions, we can explicitly construct a good approximation to the solution, the existence of which has already been guaranteed. Thus, we can investigate in detail the way the solution propagates under some particular condition for one of the phenomena governed by the equations. In Chapter 3 we look at the propagation of a particularly important wave with a high frequency.

Light is a good example of a high frequency wave. If we take a minute to consider everyday occurrences, then what stands out is that light travels in a straight line and, also, when it hits a boundary it is reflected. Another light phenomenon that probably is familiar to us from experience is the refraction of light that occurs at the boundary of air and water. Properties such as those just mentioned for light (i.e., a slew of properties of waves that are commonly, if unconsciously, observed by us) in fact, can be explained mathematically by making use of the asymptotic solution.

By similar means, when the solution has a discontinuity, we can study how this discontinuity is transmitted.

In Chapter 4, with respect to the decay of the local energy, we will investigate the propagation of a wave outside a bounded obstacle. If there is no loss of energy due to, say, friction, then the total energy for the wave does not change with time. However, the positions that bear the energy, i.e., the positions of vibration/oscillation, will gradually recede into the distance with the passage of time.

Now, suppose that an observer is standing at some particular point. With the person stationary, let us assume that a vibration or oscillation arises somewhere and subsequently this leads to the transmission of a wave. Further, this wave passes by the observer and on hitting some solid object the wave is reflected repeatedly. After it has passed by the stationary observer a number of times, the effect, with time, will begin to diminish, and eventually in the environs of the observer the waves will die out. In fact, the consideration of such a phenomenon is the same as the investigation of the decay of local energy.

The reader should keep in mind that the problems that we consider in Chapters 3 and 4 relate only to very particular characteristics of waves. In contrast to a wave with a high frequency, a wave with an extremely low frequency, that is, one with a very high wavelength, has virtually no aftereffect even if it does hit some obstacle. Thus, what we can say is that the wave passes by without any reflection. In this book, we will not consider such a phenomenon. Also, we will not cover how the obstacle's shadow effects the wave. This problem is quite a difficult one. However, from the perspectives of both mathematics and physics, this problem is a very interesting one and worthy of future research.

If we try to encapsulate the formal aims of this book, then, first, it is to prove in the strict mathematical sense the existence of a solution, and, in addition, it is to describe the basis of the mathematical means that explain the most typical properties of a wave. To aid the reader's comprehension of our fulfillment of these aims, we will on several occasions repeat similar arguments.

Nomenclature used in this book

We denote the n-dimensional Euclidean space by $\mathbb{R}^n$ and a point in $\mathbb{R}^n$ by $x = (x_1, x_2, \ldots, x_n)$ with $x_j \in \mathbb{R}$ $(j = 1, 2, \ldots, n)$.

(1) Suppose D is an open set in $\mathbb{R}^n$. Then we will write the partial derivative of first order of a function f defined on D as

$$\frac{\partial f}{\partial x_j} \quad \text{or} \quad f_{x_j}.$$

For the exponent $\alpha = (\alpha_1, \alpha_2, \ldots, \alpha_n)$ with $\alpha_j \in \{0, 1, 2, \ldots\}$ and $|\alpha| = \alpha_1 + \cdots + \alpha_n$, the partial derivatives of higher order,

$$\frac{\partial^{|\alpha|} f}{\partial x_1^{\alpha_1} \cdots \partial x_n^{\alpha_n}},$$

will be denoted by

$$\partial_x^\alpha f(x) \quad \text{or} \quad f^{(\alpha)}(x).$$

(2) We denote by $C(D)$ the space consisting of all the continuous functions that are defined on D. Then, for $m = 0, 1, 2, \ldots$ we define $C^m(D)$ by $C^0(D) = C(D)$ and otherwise by

$$C^m(D) = \{f \in C(D); \text{ for all } |\alpha| \leqslant m \quad f^{(\alpha)} \in C(D)\}.$$

Further, we set $C^\infty(D) = \bigcap_{m=1}^{\infty} C^m(D)$.

(3) We set $\mathcal{B}(D) = \{f \in C(D); f \text{ is bounded in } D\}$; then for $m = 1, 2, \ldots$ we define

$$\mathcal{B}^m(D) = \{f \in C^m(D); \text{ for all } |\alpha| \leqslant m \quad f^{(\alpha)} \in \mathcal{B}(D)\}.$$

In addition, we set $\mathcal{B}^\infty(D) = \bigcap_{m=1}^{\infty} \mathcal{B}^m(D)$.

(4) Let E be a linear space. Then $C^m(D; E)$ denotes all the functions defined on D that take their values in E and are functions with continuous partial derivatives of mth order with respect to the topology on E.

(5) We will denote Sobolev spaces by $H^m(D)$ $(m = 0, 1, 2, \ldots)$; a definition of these spaces is given in §2.2(a).

CHAPTER 1

Wave Phenomena and Hyperbolic Equations

For several of the wave phenomena that we might see or experience around us, we can derive the partial differential equations that govern their behaviour. Even though the primary physical factors that give rise to these wave phenomena differ, by examining these partial differential equations it is possible to find some common form for them. Having said this, what we call *wave phenomena* come in various types, and moreover, they are quite diverse. However, the reason that the characteristic of a wave is clearly common to these phenomena can be understood readily by looking at the form of the partial differential equations that govern the phenomena.

In the set of all partial differential equations there exists a class of partial differential equations called *equations of hyperbolic type* (or simply *hyperbolic equations*). The above-mentioned equations of wave phenomena belong to this class. Now, our intention is to introduce several fundamental concepts of this class of equations. So, since wave equations are exemplary instances of hyperbolic equations, we will derive expressions for solutions of their initial value problem, and, in addition, find some properties of these solutions.

1.1 Equations of wave phenomena

(a) The vibration of a string.

Let us suppose a string is stretched between 2 points A and B. We can assume that A and B lie along the x-axis at coordinates a and b, respectively, with $a < b$. In this context, every point of the string corresponds to a point in the interval $[a, b]$. Actually, we can restrict our considerations to strings for which the vibration is only in a vertical direction relative to the state of rest. With this in mind, we assume the string is stretched over the interval $[a, b]$ with a tension T.

Since the displacement of the string that we will examine will be small, we can also ignore the change of tension due to this displacement.

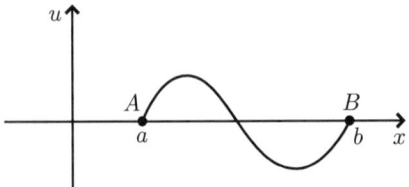

FIGURE 1.1

So, with the above in mind, at time t, we denote by $u(t, x)$ the displacement of the string at $x \in [a, b]$. If a certain part of the string is shifted from its state of rest, then the tension will produce a force that pulls back this part towards its position of rest. A consequence of this force is to impart an acceleration to this part of the string. As we shall soon see, an expression for this relationship in terms of the partial differentials of u gives us a partial differential equation.

To begin, we denote by P the point of the string that corresponds to x and by Q that which corresponds to $x + \Delta x$, and, further, we let ΔS be the infinitesimal part of the string that lies between P and Q. At time t, we can express P as $(x, u(t, x))$ and we can express Q as $(x + \Delta x, u(t, x + \Delta x))$.

Next, let us consider the force that is applied to the infinitesimal part ΔS. On ΔS at the points P and Q there is a force of strength T that is tangential to the string and in an outward direction with respect to ΔS. We shall use θ and $\theta + \Delta \theta$ to designate the angles the x-axis makes, respectively, with each of the two tangents; see also Figure 1.2.

Therefore, it is easy to see that the horizontal and vertical forces applied to ΔS are:

horizontal force: $F_h = T \cos(\theta + \Delta\theta) - T \cos\theta,$
vertical force: $F_v = T \sin(\theta + \Delta\theta) - T \sin\theta.$

However, since

$$\tan(\theta + \Delta\theta) = u_x(t, x + \Delta x) \quad \text{and} \quad \tan\theta = u_x(t, x),$$

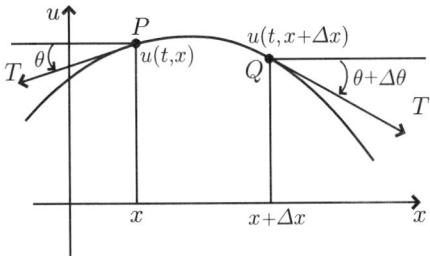

FIGURE 1.2

it follows that

$$\cos(\theta + \Delta\theta) - \cos\theta$$
$$= (1 + (u_x(t, x + \Delta x))^2)^{-1/2} - (1 + (u_x(t, x))^2)^{-1/2}$$
$$= \Delta x\, u_x(t, \xi) u_{xx}(t, \xi)(1 + (u_x(t, \xi))^2)^{-3/2}$$
$$(x < \xi < x + \Delta x)$$

and

$$\sin(\theta + \Delta\theta) - \sin\theta$$
$$= u_x(t, x + \Delta x)(1 + (u_x(t, x + \Delta x))^2)^{-1/2}$$
$$- u_x(t, x)(1 + (u_x(t, x))^2)^{-1/2}$$
$$= \Delta x\, u_{xx}(t, \eta)(1 + (u_x(t, \eta))^2)^{-3/2} \qquad (x < \eta < x + \Delta x).$$

Now, if Δx is sufficiently small, then we can take ξ and η to be sufficiently close, and hence the above formulae yield

$$F_h \doteqdot u_x(t, x) F_v.$$

Further, if the vibration is sufficiently small so that we can consider $u_x(t, x)$ also to be sufficiently small, then the force applied in the horizontal direction in comparison to that in the vertical direction is negligible. So, as noted above, we can restrict ourselves to the consideration of motion only in the vertical direction.

Next, let us denote the density of the string at the point x by $\rho(x)$. Since the mass of ΔS is $\rho(x)\Delta x$, the force that gives an acceleration $u_{tt}(t, x)$ to ΔS is $\rho(x)\Delta x u_{tt}(t, x)$. If at time t the external force per unit length applied in a vertical direction at x is $F(t, x)$, then the external force applied to ΔS is $\Delta x F(t, x)$. Hence, the forces applied to ΔS and the acceleration are connected as follows:

$$T\Delta x u_{xx}(t, \eta)(1 + (u_x(t, \eta))^2)^{-3/2} + \Delta x F(t, x) = \rho(x)\Delta x u_{tt}(t, x).$$

By dividing both sides by Δx and then by allowing $\Delta x \to 0$, since $\eta \to x$, we obtain that

$$(1.1) \qquad \rho(x) u_{tt}(t, x) = T u_{xx}(1 + (u_x(t, x))^2)^{-3/2} + F(t, x).$$

The above equation necessarily holds for all points $x \in (a, b)$.

In the above discussion, we have taken both u and u_x to be sufficiently small. Within this context, we can assume that we can expand u and F, respectively, in terms of the parameter ε, which itself is deemed to be small, as follows:

$$(1.2) \qquad u(t, x) = \varepsilon u_0(t, x) + \varepsilon^2 u_1(t, x) + \cdots$$

and

$$(1.3) \qquad F(t, x) = \varepsilon f(t, x),$$

where $u_0, u_1, \ldots$ are assumed to be smooth functions with respect to t and x. If we now substitute equations (1.2) and (1.3) into (1.1) and then equate the coefficient of ε^1 on both sides, we see that

$$(1.4) \qquad \rho(x)\frac{\partial^2 u_0}{\partial t^2}(t, x) = T\frac{\partial^2 u_0}{\partial x^2}(t, x) + f(t, x).$$

Next, by equating the coefficient of ε^2 on both sides we get

$$(1.5) \qquad \rho(x)\frac{\partial^2 u_1}{\partial t^2}(t, x) = T\frac{\partial^2 u_1}{\partial x^2}(t, x).$$

Continuing this process to ε^3 we obtain that

$$(1.6) \quad \rho(x)\frac{\partial^2 u_2}{\partial t^2}(t, x) = T\frac{\partial^2 u_2}{\partial x^2}(t, x) - \frac{3}{2}\frac{\partial^2 u_0}{\partial x^2}(t, x)\left(\frac{\partial u_0}{\partial x}(t, x)\right)^2.$$

Suppose that these equations can be solved with the *proviso* that $u_0, u_1, \ldots$ are smooth functions. Further, let us take $u(t, x)$ given in (1.2) to be a solution of (1.1). By making ε particularly small so that $\varepsilon^2 u_1 + \varepsilon^3 u_2 + \cdots$ is negligible when compared to εu_0, we can take the motion of the string to be described solely by εu_0. So, we can write $u(t, x) = \varepsilon u_0(t, x)$, and from (1.4) we see that $u(t, x)$ satisfies

$$(1.7) \qquad \rho(x)\frac{\partial^2 u(t, x)}{\partial t^2} = T\frac{\partial^2 u(t, x)}{\partial x^2} + F(t, x).$$

The partial differential equation given in (1.7) is called the 1-dimensional *wave equation* or on occasion *the equation of a string*. If we can solve this equation, then we will be able to understand the vibration of a string with small amplitude.

In general, there are many cases for which the vibration of the string also experiences a certain amount of friction which cannot readily be ignored in our calculations. For example, if the string vibrates outside a vacuum it will encounter friction from the air around it. When there is some friction, we need to include in our considerations a force that is approximately proportional in size to the velocity at that point, but in the direction opposite to the motion of the string at that point. Therefore, with the above in mind, we need to modify (1.1) to take into account the influence of the friction; namely,

$$\rho(x)u_{tt}(t,x) = Tu_{xx}(1 + (u_x(t,x))^2)^{-3/2} - \alpha u_t(t,x) + F(t,x),$$

where α is a positive constant. In the cases for which the amplitude is small, equation (1.7) will become

$$(1.8) \qquad \rho(x)u_{tt}(t,x) - Tu_{xx} + \alpha u_t(t,x) = F(t,x).$$

We will look at the effect of friction in more detail in Chapter 3.

(b) The equation of oscillation of a spring.

Suppose a string is set in motion from some initial point. With the passage of time, although the motion is along the string, at every point of the string the vibration, itself, is vertical relative to the rest state. This is exactly the type of waves we considered in the previous section; that is to say, the wave vibrates vertically relative to the direction of propagation of the wave. Thus, these waves are called *transverse waves*.

In contrast, in this section, we consider *longitudinal waves*; i.e., waves that vibrate in a parallel manner along the direction of propagation. One way of generating such waves is by attaching the 2 ends of a coiled spring, which lies on a frictionless table, to points A and B, respectively.

So, we consider exactly what happens. Let l be the length of the spring before any force is applied. Now, let us stretch the spring and attach its ends to A and B, respectively. Since, we can assume that Hooke's law is obeyed, the amount of elongation of the spring is proportional to the strength of the force applied at the ends. To be more explicit, let k denote the spring constant, then we know that if we apply a force of strength T at the ends of a spring of length d, say, the elongation s of the spring is given by

$$\frac{s}{d} = kT.$$

As in the case of a string, we can assume that each part of the spring corresponds to some point in the interval $[a, b]$. The fact that we have stretched the spring gives us that $l < b - a$. For future reference, we set

$$\alpha = \frac{l}{b - a}.$$

Before the spring is set in motion, mark by P the point on the spring that lies above x. Assume that after time t, this point is displaced by an amount $u(t, x)$. That is to say, the point P which we have marked on the spring is now at the coordinate $x + u(t, x)$.

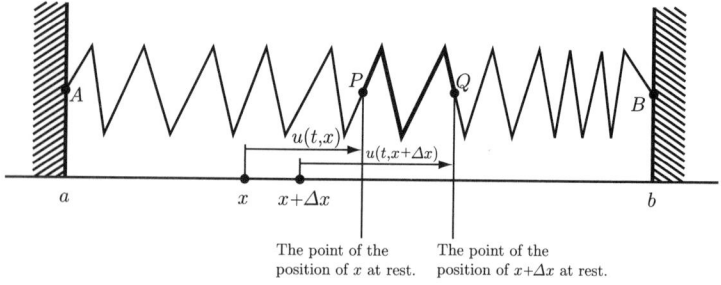

The point of the   The point of the
position of x at rest.   position of $x + \Delta x$ at rest.

FIGURE 1.3

When the spring is still in its motionless state, we denote by ΔS the infinitesimal part of the spring between x and $x + \Delta x$. Clearly, this part ΔS, in the motionless state, corresponds to the length $\Delta x \cdot l(b - a)^{-1} = \alpha \Delta x$ of the spring in its original untethered state. Now, at time t, the length of this section becomes

$$\Delta x + u(t, x + \Delta x) - u(t, x).$$

Therefore, when no force is applied, the spring is stretched by only

$$u(t, x + \Delta x) - u(t, x) + (1 - \alpha)\Delta x.$$

From Hooke's law it is easy to see that at both ends of ΔS the outward force produced is

$$k\frac{u(t, x + \Delta x) - u(t, x) + (1 - \alpha)\Delta x}{\alpha \Delta x}.$$

As $\Delta x \to 0$, the above ratio converges and the limit corresponds to the force applied at P.

From the above, it follows that "at $x + u(t, x)$ there is a force of strength

$$k\left\{\frac{1}{\alpha}\frac{\partial u}{\partial x}(t, x) + \left(\frac{1}{\alpha} - 1\right)\right\}$$

that pulls the right-side part leftwards from the left-side part and then pulls the left-side part rightwards from the right-side part".

Next, for $\Delta x > 0$, we find the difference in the force that is applied from the end points to ΔS. At P, the size of the force that is applied leftwards is given by $k(\alpha^{-1}u_x(t, x) + \alpha^{-1} - 1)$, while at the other end of ΔS, Q, there is a force in a rightward direction given by $k(\alpha^{-1}u_x(t, x + \Delta x) + \alpha^{-1} - 1)$. Therefore, as a whole in ΔS there is a force rightwards of

$$k\alpha^{-1}\{u_x(t, x+\Delta x)-u_x(t, x)\} = k\alpha^{-1}\Delta x u_{xx}(t, \xi) \quad (x < \xi < x+\Delta x).$$

Now, if we set the density of the spring when it is stretched along the interval AB to be ρ, then since the mass of ΔS is given by $\rho\Delta x$, we find that the relationship between the acceleration of this part and the force applied there is given by

$$\rho\Delta x u_{tt}(t, x) = k\alpha^{-1}\Delta x u_{xx}(t, \xi) + \Delta x F(t, x),$$

where $F(t, x)$ denotes the external force per unit length that is applied at P. Thus, if we divide both sides by Δx and let $\Delta x \to 0$, we see that

$$(1.9) \qquad \rho u_{tt}(t, x) = k\alpha^{-1}u_{xx}(t, x) + F(t, x).$$

The partial differential equation in (1.9) becomes the equation of the string in (1.7) by the simple substitution of k/α by T. Thus, even though the primary dynamical factors that give rise to the motions of a string and that of a spring are different, it is very interesting that their eventual equations have the same form.

(c) The equation for the propagation of sound.

Sound, as the propagation of motion of air, is a familiar concept. In this section, we derive the equation for the propagation of sound.

For our purposes, we will consider air to be an ideal gas. That is, suppose that D is an arbitrary domain within a gas, then the rest of the gas exerts a force (pressure) on D that is expressed in terms of a real-valued function $p(x)$, defined in the domain of the gas, and is given by

$$-\int_S p(x)\nu(x)d\sigma_x,$$

where $\nu(x) = (\nu_1(x), \nu_2(x), \nu_3(x)) \in \mathbb{R}^3$ denotes the outward unit normal vector at $x \in \partial D = S$ and $d\sigma_x$ denotes a surface element of S.

Therefore, for the x_1-coordinate the force exerted by the gas on D is

$$-\int_S p(x)\nu_1(x)d\sigma_x.$$

But, by the divergence theorem, this is equal to $-\int_D \dfrac{\partial p}{\partial x_1}(x)dx$. Since the contributions from the other directions can be dealt with in a similar fashion, it follows that the force that acts on D is

$$-\int_D \operatorname{grad} p(x)dx.$$

Now, we establish some nomenclature. At time t, for a point x in D let us denote the density by $\rho(t, x)$, the acceleration by $w(t, x) = (w_1(t, x), w_2(t, x), w_3(t, x))$, and the external force acting at this point by $f(t, x) = (f_1(t, x), f_2(t, x), f_3(t, x))$. Then, from the relationship between the acceleration and the applied forces, we have that

$$-\int_D \operatorname{grad} p(t, x)dx + \int_D f(t, x)dx = \int_D \rho(t, x)w(t, x)dx.$$

Since the above formula holds for any domain D, we can deduce that relation (1.10), below, holds for every (t, x).

(1.10) $$\rho(t, x)w(t, x) = -\operatorname{grad} p(t, x) + f(t, x).$$

Next, we express the velocity vector of the gas at each point by $v(t, x) = (v_1(t, x), v_2(t, x), v_3(t, x))$. Then, the gas, which was at the point x at time t, moves after an interval Δt to the point $x + \Delta t v(t, x)$. Therefore, the velocity of the gas, which was at point x and at time t, after an interval Δt is $v(t + \Delta t, x + \Delta t v(t, x))$. Thus, the acceleration of the gas is the limit as $\Delta t \to 0$ of

$$\frac{1}{\Delta t}(v(t + \Delta t, x + \Delta t v(t, x)) - v(t, x)).$$

Therefore, we may write

(1.11)
$$w(t, x) = \frac{\partial v}{\partial t}(t, x) + \frac{\partial v}{\partial x_1}(t, x)v_1(t, x) + \frac{\partial v}{\partial x_2}(t, x)v_2(t, x)$$
$$+ \frac{\partial v}{\partial x_3}(t, x)v_3(t, x).$$

Having obtained the above, our next objective is to study the relationship between the velocity v and the density. With this in mind, let D be an arbitrary domain with boundary S. Since the mass of the gas inside D is

$$\int_D \rho(t, x)dx,$$

the rate of change of the mass inside D in unit time is

$$\int_D \frac{\partial \rho}{\partial t}(t, x)dx.$$

On the other hand, the mass of the gas that flows out from D through the boundary S in unit time is

$$\int_S \rho(t, x)v(t, x) \cdot \nu(x)d\sigma_x.$$

Again with recourse to the divergence theorem, this can be expressed as

$$\int_D \left\{ \frac{\partial}{\partial x_1}(\rho(t, x)v_1(t, x)) + \frac{\partial}{\partial x_2}(\rho(t, x)v_2(t, x)) \right.$$
$$\left. + \frac{\partial}{\partial x_3}(\rho(t, x)v_3(t, x)) \right\}dx.$$

Therefore, we have that

$$\int_D \frac{\partial \rho}{\partial t}(t, x)dx = -\int_D \sum_{j=1}^3 \frac{\partial}{\partial x_j}(\rho(t, x)v_j(t, x))dx.$$

The above equation holds for any arbitrary domain D, so

$$(1.12) \qquad \frac{\partial \rho}{\partial t}(t, x) + \operatorname{div}(\rho(t, x)v(t, x)) = 0.$$

This equation is usually called the *equation of continuity*.

Now, let ρ_0 be the uniform density of the air at rest, and, further, let the function $u(t, x)$ denote the rate of divergence from ρ_0. This latter function expresses the compression rate of the gas and can be defined from

$$\rho(t, x) = \rho_0(1 + u(t, x)).$$

Now we assume the relationship between the density and pressure of the gas is described by the following: "the pressure and density are proportional"; that is to say, there exists a constant $K > 0$ such that

$$(1.13) \qquad p(t, x) = K\rho(t, x) = K\rho_0(1 + u(t, x)),$$

the constant K is usually called the *bulk modulus*. Thus, we have that

$$(1.14) \qquad \operatorname{grad} p = K\rho_0 \operatorname{grad} u.$$

We wish to assume that the gas has only slight motion. Let us be more precise. To begin with, for a sufficiently small $\varepsilon > 0$, let the velocity vector v, the compression rate u, and the external force f have, respectively, the following forms,

$$v = \varepsilon v_0 + \varepsilon^2 v_2 + \cdots,$$
$$u = \varepsilon u_0 + \varepsilon^2 u_1 + \cdots,$$
$$f = \varepsilon f_0.$$

From (1.10) and (1.11) and with the aid of (1.14), comparing the coefficients of the powers of ε, we obtain that

$$(1.15) \qquad \rho_0 \frac{\partial v_0}{\partial t} = -K\rho_0 \operatorname{grad} u_0 + f_0.$$

Also, from (1.12) we have that

$$(1.16) \qquad \rho_0 \frac{\partial u_0}{\partial t} + \rho_0 \operatorname{div} v_0 = 0.$$

Next, by taking the "div" of both sides of (1.15), we obtain

$$\rho_0 \frac{\partial}{\partial t} \operatorname{div} v_0 = -K\rho_0 \operatorname{div} \operatorname{grad} u_0 + \operatorname{div} f_0.$$

Now, by applying the partial derivative with respect to t to both sides of (1.16), we obtain

$$\rho_0 \frac{\partial^2 u_0}{\partial t^2} + \rho_0 \frac{\partial}{\partial t} \operatorname{div} \boldsymbol{v}_0 = 0.$$

By suitably combining the latter two equations, we see that

$$\rho_0 \frac{\partial^2 u_0}{\partial t^2} - K \rho_0 \operatorname{div} \operatorname{grad} u_0 + \operatorname{div} \boldsymbol{f}_0 = 0.$$

But, since $\operatorname{div} \operatorname{grad} u_0 = \Delta u_0$, the following equation holds:

(1.17) $$\frac{\partial^2 u_0}{\partial t^2} - K \Delta u_0 = f_0 \qquad \left(f_0 = -\frac{1}{\rho_0} \operatorname{div} \boldsymbol{f}_0 \right).$$

Thus, if we assume that u_0 can be determined by (1.17), then we can find $\boldsymbol{v}_0$ by means of (1.15). Continuing, if equation (1.17) is solvable for an arbitrary right-hand side, we can find, in order, u_j, $\boldsymbol{v}_j$ $(j = 1, 2, \ldots)$. From the above, for the propagation of a sound wave with a small amplitude, we can assume that the terms in ε^2 and higher are negligible. Hence, it follows from (1.17) that u is governed by

(1.18) $$\frac{\partial^2 u}{\partial t^2} - K \Delta u = f \qquad \left(f = -\frac{1}{\rho_0} \operatorname{div} \boldsymbol{f} \right).$$

By calling on Theorem 1.2, which is discussed later in this chapter, or on our deliberations in Chapter 3, we can show that the coefficient K that appears in equation (1.18) determines the speed of propagation of the sound wave. To be precise, $\sqrt{K}$ is the speed of propagation of the sound wave.

From (1.13), the bulk modulus K is a constant of the proportional relationship between the pressure and density of the gas, and we can treat K as a constant that does not depend on place and time. This latter assumption poses no intrinsic problem for us as long as the region of the propagation of sound is relatively short. However, there are many cases where a wave traverses a relatively long distance, and in such cases K can no longer be treated as a constant.

As noted above, K is the constant of proportionality between the pressure and density of the gas, but we need to say that K is not independent of the temperature of the gas. In fact, within the confines of an ideal gas, K is proportional to the absolute temperature.

Therefore, if we assume the temperature of the gas changes in accordance with its position within the gas, then K becomes a function in x. Thus, for these cases, we need to change (1.18) to the following:

$$(1.19) \qquad \frac{\partial^2 u}{\partial t^2} - \sum_{j=1}^{3} \frac{\partial}{\partial x_j} \left(K(x) \frac{\partial u}{\partial x_j} \right) = f.$$

The fact that $K(x)$ is no longer a constant causes the phenomena (of wave propagation) governed by (1.19) to move along a curve dependent on $K(x)$ rather than along a straight line. With regard to the propagation of sound, in cases when the earth's surface is cold and the atmosphere is warm, it is theoretically possible, and perhaps you have experienced this, to hear a far distant sound. This phenomenon is the result of the change of $K(x)$ depending on position in the air. We will explain this phenomenon as a property of the solution of (1.19) in Chapter 3, Section 3(c).

(d) Maxwell's equations, elastic equations.

Up to this point, we have concerned ourselves with equations in which there is a solitary unknown function, with values in $\mathbb{R}$ or $\mathbb{C}$. Thus, in each case there is only a single equation. In the sequel, we will look at equations with several unknown functions. Obviously, since there are several unknown functions the equations must satisfy several relations, we will call such a situation a *system of equations*. Below, we will investigate two sorts of systems of equations that describe wave phenomena; namely, Maxwell's equations and elastic equations.

Maxwell's equations.

Succinctly, Maxwell's equations are equations that describe the phenomenon of electromagnetism. Any good primer in undergraduate physics should provide the background and nomenclature for Maxwell's equations. Thus, to avoid needless repetition, we shall only state the results.

At time t and for $x \in \mathbb{R}^3$ we denote the electrical field by $\boldsymbol{E}(t, x)$ and the magnetic field by $\boldsymbol{B}(t, x)$. The electrical and magnetic fields can be expressed as elements of $\mathbb{R}^3$. That is, we can write $\boldsymbol{E}(t, x)$ as $(E_1(t, x), E_2(t, x), E_3(t, x))$ and $\boldsymbol{B}(t, x)$ as $(B_1(t, x), B_2(t, x), B_3(t, x))$ with the real-valued functions $E_j(t, x)$ and $B_j(t, x)$ $(j = 1, 2, 3)$. Further, we assume that at time t and at the point x the electric charge density is $q(t, x)$ and the flow of electric charge is $\boldsymbol{j}(t, x)$.

Using this notation, we can write that these quantities satisfy the following equations:

$$(1.20) \qquad \varepsilon_0 \mathrm{div}\, \boldsymbol{E} = q,$$

$$(1.21) \qquad \mathrm{rot}\, \boldsymbol{E} = -\frac{\partial \boldsymbol{B}}{\partial t},$$

$$(1.22) \qquad \mathrm{div}\, \boldsymbol{B} = 0,$$

$$(1.23) \qquad \frac{1}{\mu_0} \mathrm{rot}\, \boldsymbol{B} = \varepsilon_0 \frac{\partial \boldsymbol{E}}{\partial t} + \boldsymbol{j},$$

where ε_0 and μ_0, respectively, are the permittivity and permeability of a vacuum. This system of equations is commonly referred to as *Maxwell's equations*.

By considering q and $\boldsymbol{j}$ as given functions, each component of the electric field $\boldsymbol{E}$ and each component of the magnetic field $\boldsymbol{B}$, in total, form a collection of six unknown functions.

With regard to $\boldsymbol{E}$ and $\boldsymbol{B}$ that satisfy Maxwell's equations, it is known that there exists a pair $(\boldsymbol{A}, \varphi)$, called the Lorentz gauge, consisting of a $\mathbb{R}^3$-valued function $\boldsymbol{A}$ and a real-valued function φ that satisfy the following:

$$(1.24) \qquad \boldsymbol{B} = \mathrm{rot}\, \boldsymbol{A},$$

$$(1.25) \qquad \boldsymbol{E} = -\mathrm{grad}\, \varphi - \frac{\partial \boldsymbol{A}}{\partial t},$$

$$(1.26) \qquad \varepsilon_0 \frac{\partial \varphi}{\partial t} + \frac{1}{\mu_0} \mathrm{div}\, \boldsymbol{A} = 0.$$

Using the above, we can confirm that Maxwell's equations are in fact wave equations.

First substitute (1.25) into the left-hand side of (1.20) to obtain that

$$(1.27) \qquad -\varepsilon_0 \mathrm{div}\, \mathrm{grad}\, \varphi - \varepsilon_0 \mathrm{div}\, \frac{\partial \boldsymbol{A}}{\partial t} = q.$$

Next, substitute (1.24) and (1.25) into the left-hand and right-hand sides, respectively, of (1.23), thus,

$$(1.28) \qquad \frac{1}{\mu_0} \mathrm{rot}\, \mathrm{rot}\, \boldsymbol{A} = -\varepsilon_0 \frac{\partial^2 \boldsymbol{A}}{\partial t^2} - \varepsilon_0 \mathrm{grad}\, \frac{\partial \varphi}{\partial t} + \boldsymbol{j}.$$

Now, by using (1.26) in (1.27), we see that

$$-\varepsilon_0 \mathrm{div}\,\mathrm{grad}\,\varphi + \varepsilon_0^2 \mu_0 \frac{\partial^2 \varphi}{\partial t^2} = q.$$

But, since $\mathrm{div}\,\mathrm{grad} = \Delta$, we have that

$$(1.29) \qquad \varepsilon_0 \mu_0 \frac{\partial^2 \varphi}{\partial t^2} - \Delta \varphi = \frac{q}{\varepsilon_0}.$$

Finally, we can eliminate φ in favour of $\boldsymbol{A}$. This is done as follows; first, we take the "grad" of both sides of (1.26) and then in this derived equation we use (1.28). Now, by noting that $\mathrm{grad}\,\mathrm{div} - \mathrm{rot}\,\mathrm{rot} = \Delta$, our manipulations yield that

$$(1.30) \qquad \varepsilon_0 \mu_0 \frac{\partial^2 \boldsymbol{A}}{\partial t^2} - \Delta \boldsymbol{A} = \mu_0 \boldsymbol{j}.$$

Conversely, suppose that $(\boldsymbol{A}, \varphi)$ satisfies (1.29) and (1.30) and further that the relation in (1.26) holds. Then it is easy to show that $\boldsymbol{B}$ and $\boldsymbol{E}$ as given in (1.24) and (1.25), respectively, satisfy the formulae (1.20)–(1.23).

Thus, using the Lorentz gauge $(\boldsymbol{A}, \varphi)$ it is possible to confirm that Maxwell's equations are such that φ and each component of $\boldsymbol{A}$ satisfy, respectively, the wave equation and, moreover, they also satisfy (1.26).

Elastic equations.

By applying a force to an object suppose that we cause a change in its shape and volume. Subsequently, on removing the force, suppose the object regains it original shape. This property is called *elasticity,* while the object, itself, is said to be *elastic.* It is very enlightening to study the propagation phenomena of a wave in such an elastic object; as might be expected, such a wave is called an *elastic wave.* For instance, the external shell of the earth, namely, the earth's crust, is elastic, and so an earthquake is just an elastic wave transmitted through the earth's crust.

Our intention is to introduce an equation that describes evenly in all directions the propagation of a small amplitude wave in a uniform elastic body. With this in mind, we assume that Ω, a region in $\mathbb{R}^3$, is part of an elastic object that, at present, is in a state of rest and to which no force is applied. Further, from its position x at rest, suppose that at time t the point x is displaced by an amount $\boldsymbol{u}(t, x)$ from x. In other words, from its original point x it moves to the point

$x + \boldsymbol{u}(t, x)$. Since $\boldsymbol{u}(t, x)$ is an element of $\mathbb{R}^3$ we can express $\boldsymbol{u}(t, x)$ in the form $\boldsymbol{u}(t, x) = (u_1(t, x), u_2(t, x), u_3(t, x))$. It is known that the vector function $\boldsymbol{u}(t, x)$ that describes this shift satisfies the following equation, usually called the *elastic equation*,

$$(1.31) \qquad \rho \frac{\partial^2 \boldsymbol{u}}{\partial t^2} = \alpha \Delta \boldsymbol{u} + \beta \operatorname{grad} \operatorname{div} \boldsymbol{u} + \boldsymbol{f},$$

where the function $\boldsymbol{f}(t, x) = (f_1(t, x), f_2(t, x), f_3(t, x))$ represents the external force applied at time t to the part at position x of the rest state. Further, ρ is a constant that denotes the density of the elastic object. Finally, α and β are positive constants which depend on the constants μ and λ in the following way, $\alpha = \mu$ and $\beta = \mu + \lambda$, where μ and λ are constants, usually referred to as *Lamé constants*, that denote characteristics of the elastic body.

Now, (1.31) is a system of equations with 3 unknown functions. For if we consider both sides of (1.31) for each of the components, we may collectively write

$$(1.32) \quad \rho \frac{\partial^2 u_l}{\partial t^2} = \alpha \sum_{j=1}^{3} \frac{\partial^2 u_l}{\partial x_j^2} + \beta \frac{\partial}{\partial x_l} \left(\sum_{p=1}^{3} \frac{\partial u_p}{\partial x_p} \right) + f_l \qquad (l = 1, 2, 3).$$

Sometimes in reports of earthquakes one might hear or read the terms *P*-wave and *S*-wave. We quickly explain the nature of these waves.

For earthquakes in areas of uniform geology, the propagation of the quake is governed by the elastic equation in (1.31). In such cases, it follows from (1.31) that there exist two types of waves with different properties. One is termed a longitudinal wave and has propagation speed $\sqrt{(\alpha + \beta)/\rho}$; for such waves the direction of motion and the direction of propagation of the wave are the same. The other wave is termed a transverse wave and has propagation speed $\sqrt{\alpha/\rho}$; in contrast to a transverse wave, the direction of motion and the direction of propagation of the wave are perpendicular to each other. Since the propagation speed of the longitudinal wave is greater than that of the transverse wave, the longitudinal wave arrives first at a position; i.e., the longitudinal wave is the primary wave and so, oftentimes, is termed a *P-wave*, while the transverse wave comes second, and so is usually termed an *S-wave*.

In Chapter 3 we will again consider this phenomenon but from the point of view of an asymptotic solution.

1.2 Hyperbolic partial differential operators

The particular class of partial differential operators said to be of *hyperbolic type* contains the equations and the system of equations that govern several of the wave phenomena given so far. Here, we will only concern ourselves with second-order partial differential operators and tailor the definition of hyperbolic type appropriately.

(a) Hyperbolic differential operators.

Consider the following $N \times N$ matrix,

$$A_{jl}(t,x) \quad (j,l = 1, 2, \ldots, n), \qquad H_j(t,x) \quad (j = 0, 1, \ldots, n),$$
$$A_j(t,x) \quad (j = 1, 2, \ldots, n), \qquad A_0(t,x),$$

whose entries are elements of $\mathcal{B}^\infty(\mathbb{R} \times \mathbb{R}^n)$.

Further, suppose that $u_j(t,x)$ $(j = 1, 2, \ldots, N)$ are unknown functions, and set $\boldsymbol{u}(t,x) = {}^t(u_1(t,x), u_2(t,x), \ldots, u_N(t,x))$. Finally, consider the partial differential operator P, given below, that acts on $\boldsymbol{u}$.

(1.33)
$$
\begin{aligned}
P\boldsymbol{u} &= \frac{\partial^2 \boldsymbol{u}}{\partial t^2} + \sum_{j,l=1}^{n} A_{j,l} \frac{\partial^2 \boldsymbol{u}}{\partial x_j \partial x_l} + \sum_{j=1}^{n} A_j \frac{\partial \boldsymbol{u}}{\partial x_j} \\
&\quad + \sum_{j=1}^{n} 2H_j \frac{\partial^2 \boldsymbol{u}}{\partial x_j \partial t} + H_0 \frac{\partial \boldsymbol{u}}{\partial t} + A_0 \boldsymbol{u}.
\end{aligned}
$$

If we look back at the wave equation in (1.18), then we may recover the equation from the above by setting $n = 3$, $N = 1$,

$$A_{jl}(t,x) = -K\delta_{jl},$$

and setting all of H_j, A_j and A_0 to zero. In the above, δ_{jl} is the usual Kronecker delta, namely,

$$
\delta_{jl} = \begin{cases} 1, & j = l, \\ 0, & j \neq l; \end{cases}
$$

while, in the case of the elastic equation given in (1.31), we need to take $n = 3$ and $N = 3$. Although we do not explicitly write down the coefficient matrices, we note that from (1.32) it is easy to explicitly determine the A_{jl}.

For a second-order partial differential operator P, we will say that the *principal part* of P is

$$(1.34) \qquad P_0 = \frac{\partial^2}{\partial t^2} I_N + 2 \sum_{j=1}^n H_j \frac{\partial^2}{\partial x_j \partial t} + \sum_{j,l=1}^n A_{jl} \frac{\partial^2}{\partial x_j \partial x_l}.$$

For $\lambda \in \mathbb{C}$ and $\xi = (\xi_1, \xi_2, \ldots, \xi_n) \in \mathbb{R}^n$, let
(1.35)

$$p_0(t, x, \lambda, \xi) = \det \left(\lambda^2 I_N + 2 \sum_{j=1}^n H_j(t,x)\xi_j \lambda + \sum_{j,l=1}^n A_{jl}(t,x)\xi_j \xi_l \right).$$

$p_0(t, x, \lambda, \xi)$ is called the *characteristic polynomial* of the partial differential operator P. If we consider $(t, x, \xi) \in \mathbb{R} \times \mathbb{R}^n \times \mathbb{R}^n$ to be a parameter and p_0 to be a polynomial in λ, then the degree of the characteristic polynomial is $2N$. We denote the roots of $p_0(t, x, \lambda, \xi) = 0$ by $\lambda_k(t, x, \xi)$ $(k = 1, 2, \ldots, 2N)$; we call these roots the *characteristic roots* of the partial differential operator P.

DEFINITION 1.1. The second-order partial differential operator P given in (1.33) is said to be of *hyperbolic type* (or simply just *hyperbolic*) in the t-direction if for an arbitrary parameter (t, x, ξ) the characteristic roots of P, $\lambda_k(t, x, \xi)$ $(k = 1, 2, \ldots, 2N)$, are all real. Further, if

$$\inf |\lambda_k(t, x, \xi) - \lambda_j(t, x, \xi)| > 0$$

for $k \neq j$, then P is said to be *regularly hyperbolic*. In the above, we assume that the infinum is taken over $(t, x) \in \mathbb{R} \times \mathbb{R}^n$ and $|\xi| = 1$.

The characteristic polynomial of the wave equation in (1.19) is given by

$$(1.36) \qquad p_0(\lambda, \xi) = \lambda^2 - K|\xi|^2.$$

Hence, there are 2 characteristic roots, $\sqrt{K}|\xi|$ and $-\sqrt{K}|\xi|$.

In the case of the elastic equation, we have
(1.37)
$$p_0(\lambda, \xi)$$
$$= \det \begin{bmatrix} \lambda^2 - \rho^{-1}(\alpha|\xi|^2 + \beta\xi_1^2) & -\rho^{-1}\beta\xi_1\xi_2 & -\rho^{-1}\beta\xi_1\xi_3 \\ -\rho^{-1}\beta\xi_1\xi_2 & \lambda^2 - \rho^{-1}(\alpha|\xi|^2 + \beta\xi_2^2) & -\rho^{-1}\beta\xi_2\xi_3 \\ -\rho^{-1}\beta\xi_1\xi_3 & -\rho^{-1}\beta\xi_2\xi_3 & \lambda^2 - \rho^{-1}(\alpha|\xi|^2 + \beta\xi_3^2) \end{bmatrix}$$
$$= (\lambda^2 - \rho^{-1}(\alpha + \beta)|\xi|^2)(\lambda^2 - \rho^{-1}\alpha|\xi|^2)^2.$$

Roughly speaking, from the characteristic equation one can deduce that the elastic equation contains three phenomena corresponding to the wave equation in (1.18). We shall discuss this in more detail in Chapter 3.

Characteristic surface.

In a region $U \subset \mathbb{R} \times \mathbb{R}^n$, let us suppose that we can find a real-valued function $\Phi(t, x)$ with continuous partial derivatives of first order such that $(\Phi_t, \Phi_x) \neq 0$ and the following is satisfied:

$$(1.38) \qquad p_0(t, x, \Phi_t(t, x), \Phi_x(t, x)) = 0, \qquad (t, x) \in U.$$

The set of (t, x) that satisfies $\Phi(t, x) = $ *a constant* is a C^1 surface. The surface with $\Phi(t, x) = $ *a constant* is often called a *characteristic surface*.

With regard to the wave equation (1.18), since $p_0 = \lambda^2 - v^2 |\xi|^2$ and $v = \sqrt{K}$, with $\xi \in \mathbb{R}^3$ and $|\xi| = 1$, if we set

$$\Phi(t, x) = vt + x \cdot \xi,$$

then Φ satisfies (1.38), and for a constant c the plane given by $vt + x \cdot \xi = c$, namely,

$$(1.39) \qquad \{(t, x); x \cdot \xi = c - vt\},$$

is a characteristic surface.

We will deal in more detail in Chapter 3 with the role played by characteristic surfaces. For the time being, we simply say, "The most important factors of a solution lie on the characteristic surface." For example, if the solution has a discontinuity, then this discontinuity is also exhibited by the characteristic surface. That is, to see how discontinuous surfaces move in space with respect to time, it is extremely beneficial to understand the movement of the section of the characteristic surface due to the time variable.

(b) Problems that we will consider.

For the system of hyperbolic operators of the type in (1.33), we will consider in this book the following two types of problem.

Initial value problem.

Suppose the region of space under consideration is the whole of $\mathbb{R}^n$, and hence, in this case, the space has no boundary. In this case, the problem boils down to seeking a solution for a problem for which we know the state at time $t = 0$. In other words, the problem is to find a $u(t, x)$ that satisfies

$$\begin{cases} Pu(t, x) = f(t, x) & ((t, x) \in (0, \infty) \times \mathbb{R}^n), \\ u(0, x) = u_0(x) & (x \in \mathbb{R}^n), \\ \dfrac{\partial u}{\partial t}(0, x) = u_1(x) & (x \in \mathbb{R}^n), \end{cases}$$

where $f(t, x)$ is a given $\mathbb{C}^N$-valued function in $(0, \infty) \times \mathbb{R}^n$ and $u_0(x)$, $u_1(x)$ are given $\mathbb{C}^N$-valued functions in $\mathbb{R}^n$.

Initial boundary value problem.

For the propagation of a wave there are cases where the region the wave traverses has a boundary, and in most cases of this type we need to consider carefully the existence of this boundary. These types of problems are usually called *initial boundary value problems*.

Suppose that Γ is the boundary of $\Omega \subset \mathbb{R}^n$. Then, an initial boundary value problem is the problem of finding a $u(t, x)$ for given data f, g and u_0, u_1 that satisfy

$$\begin{cases} Pu(t, x) = f(t, x) & ((t, x) \in (0, \infty) \times \Omega), \\ Bu(t, x) = g(t, x) & ((t, x) \in (0, \infty) \times \Gamma), \\ u(0, x) = u_0(x), \quad \dfrac{\partial u}{\partial t}(0, x) = u_1(x) & (x \in \Omega), \end{cases}$$

where $f(t, x)$ is a $\mathbb{C}^N$-valued function in $(0, \infty) \times \Omega$, $g(t, x)$ is a $\mathbb{C}^M$-valued function in $(0, \infty) \times \Gamma$, and $u_0(x)$, $u_1(x)$ are $\mathbb{C}^N$-valued functions in Ω. Further, B is an $M \times N$ matrix of differential operators defined on a neighbourhood of the boundary. For the particular problem under consideration, M and B are determined by the properties that the solution satisfies on the boundary.

1.3 Formulae for solutions of an initial value problem for a wave equation

In this section, we give expressions for solutions of the initial value problem for the wave equation given by

$$(1.40) \quad \begin{cases} \dfrac{\partial^2 u}{\partial t^2} - v^2 \Delta u = f(t, x) & ((t, x) \in (0, \infty) \times \mathbb{R}^n), \\ u(0, x) = \varphi(x), \quad \dfrac{\partial u}{\partial t}(0, x) = \psi(x) & (x \in \mathbb{R}^n), \end{cases}$$

where v is a positive constant.

Among hyperbolic operators, the wave equation gives rise to the most representative operator of this class, and so, the properties of its solutions will elucidate, in general, the basic properties of hyperbolic equations. Therefore, first of all, we will find expressions for solutions of this problem, and then investigate the properties of these solutions of the wave equation. Before going any further, we point out that we will restrict our considerations only to the cases of 2-dimensional and 3-dimensional spaces.

THEOREM 1.2. *For $f \equiv 0$, the solution $u(t,x)$ of the problem in (1.40) is given below. Here, we assume that $\varphi \in C^3(\mathbb{R}^n)$ and $\psi \in C^2(\mathbb{R}^n)$.*

The 3-dimensional space case. For $t > 0$, we have that
(1.41)

$$u(t,x) = \frac{\partial}{\partial t}\left(\frac{1}{4\pi v^2 t}\int_{|y-x|=vt}\varphi(y)dS_y\right) + \frac{1}{4\pi v^2 t}\int_{|y-x|=vt}\psi(y)dS_y.$$

The 2-dimensional space case.

(1.42)
$$u(t,x) = \frac{\partial}{\partial t}\left(\frac{1}{2\pi v^2 t}\int_{|x-y|<vt}\frac{\varphi(y)}{\sqrt{v^2 t^2 - |y-x|^2}}dy\right)$$
$$+ \frac{1}{2\pi v^2 t}\int_{|x-y|<vt}\frac{\psi(y)}{\sqrt{v^2 t^2 - |y-x|^2}}dy.$$

$\square$

The solution in (1.41) is known as *Kirchhoff's solution*, and the solution in (1.42) is known as *Poisson's solution*. By a change in the variable t, we show that the problem resolves itself into the case $v = 1$. To avoid cumbersome notation, we will denote the partial differential operator $\partial_t^2 - \Delta$ by $\square$, this operator is usually called the *d'Alembert operator* (or simply the *d'Alembertian*).

To begin with, we assume $u(t,x)$ is the solution of (1.40) and then set $\widetilde{u}(t,x) = u(t/v, x)$. We obtain that

$$\frac{\partial\widetilde{u}}{\partial t}(t,x) = \frac{1}{v}\frac{\partial u}{\partial t}\left(\frac{t}{v},x\right), \quad \frac{\partial^2\widetilde{u}}{\partial t^2}(t,x) = \frac{1}{v^2}\frac{\partial^2 u}{\partial t^2}\left(\frac{t}{v},x\right).$$

Hence, we see that

$$\frac{\partial^2\widetilde{u}}{\partial t^2}(t,x) - \Delta\widetilde{u}(t,x) = \left(\frac{1}{v^2}\frac{\partial^2 u}{\partial t^2} - \Delta u\right)\left(\frac{t}{v},x\right) = \frac{1}{v^2}f\left(\frac{t}{v},x\right).$$

Now, as $t \to 0$, the following hold:

$$\widetilde{u}(t,x) = u\left(\frac{t}{v}, x\right) \longrightarrow \varphi(x),$$

$$\frac{\partial \widetilde{u}}{\partial t}(t,x) = \frac{1}{v}\left(\frac{\partial u}{\partial t}\right)\left(\frac{t}{v}, x\right) \longrightarrow \frac{1}{v}\psi(x).$$

From the above, if we can find a solution that satisfies

$$\begin{cases} \left(\dfrac{\partial^2}{\partial t^2} - \Delta\right)\widetilde{u}(t,x) = \dfrac{1}{v^2}f\left(\dfrac{t}{v}, x\right) & \text{in } (0,\infty) \times \mathbb{R}^n, \\ \widetilde{u}(0,x) = \varphi(x), \quad \dfrac{\partial \widetilde{u}}{\partial t}(0,x) = \dfrac{1}{v}\psi(x), \end{cases}$$

then the solution $u(t,x)$ becomes just a matter of setting $u(t,x) = \widetilde{u}(vt, x)$.

So, our task now becomes one of using the above to confirm that (1.41) and (1.42) satisfy (1.40). But before that, a natural question to ask is, "Besides these solutions, does there exist any other solution?" The answer is no, but an exact analysis is left to our later deliberations.

The 3-dimensional space case.

First consider the case $\varphi \equiv 0$ and start with the change of variable

$$y = x + t\xi, \qquad \xi \in S^2 = \{\xi \in \mathbb{R}^3; |\xi| = 1\},$$

which leads to

(1.43) $$u(t,x) = \frac{1}{4\pi}t\int_{S^2}\psi(x + t\xi)dS_\xi.$$

Now, if we partially differentiate twice with respect to x_j, we obtain

$$\frac{\partial u}{\partial x_j}(t,x) = \frac{1}{4\pi}t\int_{S^2}\frac{\partial \psi}{\partial x_j}(x + t\xi)dS_\xi,$$

$$\frac{\partial^2 u}{\partial x_j^2}(t,x) = \frac{1}{4\pi}t\int_{S^2}\frac{\partial^2 \psi}{\partial x_j^2}(x + t\xi)dS_\xi.$$

Therefore, we have that

(1.44)
$$\begin{aligned} \Delta u(t,x) &= \frac{1}{4\pi}t\int_{S^2}(\Delta\psi)(x + t\xi)dS_\xi \\ &= \frac{1}{4\pi}\int_{|y-x|=t}(\Delta\psi)(y)dS_y. \end{aligned}$$

On the other hand, if we partially differentiate with respect to t, we get
(1.45)
$$\frac{\partial u}{\partial t}(t,x) = \frac{1}{4\pi}\int_{S^2}\psi(x+t\xi)dS_\xi + \frac{1}{4\pi}t\int_{S^2}\sum_{j=1}^{3}\frac{\partial\psi}{\partial x_j}(x+t\xi)\xi_j dS_\xi.$$

In the right-most expression of the right-hand side of the above formula, $\xi = (\xi_1, \xi_2, \xi_3) \in S^2$ is an outward unit normal vector at $y = x + t\xi$ on the sphere $\{y; |y-x| = t\}$. Hence, we may write

$$\frac{1}{4\pi}\int_{S^2}\sum_{j=1}^{3}\frac{\partial\psi}{\partial x_j}(x+t\xi)\xi_j dS_\xi = \frac{1}{4\pi t^2}\int_{|y-x|=t}\sum_{j=1}^{3}\frac{\partial\psi}{\partial y_j}(y)\nu_j(y)dS_y.$$

But, $\nu(y) = (\nu_1(y), \nu_2(y), \nu_3(y))$ is the outward unit normal vector at y on the sphere. So, by using the divergence formula on the right-hand side, we obtain that

(1.46) $$\frac{1}{4\pi}\int_{S^2}\sum_{j=1}^{3}\frac{\partial\psi}{\partial x_j}(x+t\xi)\xi_j dS_\xi = \frac{1}{4\pi t^2}\int_{|y-x|<t}(\Delta\psi)(y)dy.$$

The above now leads to the following expression:

$$\frac{\partial u}{\partial t}(t,x) = \frac{1}{4\pi}\int_{S^2}\psi(x+t\xi)dS_\xi + \frac{1}{4\pi t}\int_{|y-x|<t}(\Delta\psi)(y)dy.$$

By partially differentiating both sides with respect to t, we see that

$$\frac{\partial^2 u}{\partial t^2}(t,x) = \frac{1}{4\pi}\int_{S^2}\sum_{j=1}^{3}\frac{\partial\psi}{\partial x_j}(x+t\xi)\xi_j dS_\xi$$

$$- \frac{1}{4\pi t^2}\int_{|y-x|<t}(\Delta\psi)(y)dy + \frac{1}{4\pi t}\int_{|y-x|=t}(\Delta\psi)(y)dS_y.$$

Now, by substituting (1.46) into the first term of the right-hand side of the above and cancelling, we get

$$\frac{\partial^2 u}{\partial t^2}(t,x) = \frac{1}{4\pi t}\int_{|y-x|=t}(\Delta\psi)(y)dS_y.$$

Combining the above with (1.44) yields, as required,

$$\frac{\partial^2 u}{\partial t^2}(t,x) = \Delta u(t,x).$$

Next, we investigate the initial conditions. First, suppose that in (1.43) we let $t \to 0$, then $u(t,x) \to 0$. Second, let $t \to 0$ in (1.45), in this case we see that

$$\frac{\partial u}{\partial t}(t,x) \longrightarrow \frac{1}{4\pi} \int_{S^2} \psi(x) dS_\xi = \psi(x).$$

Note that the latter two statements show that the initial conditions are satisfied.

Now we turn our attention to the case $\varphi \not\equiv 0$. As a starting point, define

$$(1.47) \qquad M[g](t,x) = \frac{t}{4\pi} \int_{S^2} g(x + t\xi) dS_\xi \qquad (t > 0),$$

where g is a function defined on $\mathbb{R}^3$.

It follows immediately from the definition that if $g \in C^m(\mathbb{R}^3)$, then $M[g] \in C^m((0,\infty) \times \mathbb{R}^3)$. Therefore, if $\varphi \in C^3(\mathbb{R}^2)$, then $M[\varphi] \in C^3((0,\infty) \times \mathbb{R}^3)$.

In a similar way to the steps shown above, we can show that

$$(1.48) \qquad \begin{cases} \left(\dfrac{\partial^2}{\partial t^2} - \Delta\right) M[\varphi] = 0, \\[2mm] M[\varphi](t,x) \longrightarrow 0, \qquad \dfrac{\partial}{\partial t} M[\varphi](t,x) \longrightarrow \varphi(x). \end{cases}$$

The convergence, in the above, is generalized and uniform in $\mathbb{R}^3$.

Now, since $M[\varphi] \in C^3((0,\infty) \times \mathbb{R}^3)$, the following holds:

$$\left(\frac{\partial^2}{\partial t^2} - \Delta\right) \frac{\partial}{\partial t} M[\varphi] = \frac{\partial}{\partial t}\left(\frac{\partial^2}{\partial t^2} - \Delta\right) M[\varphi] = 0.$$

Hence, $v = \dfrac{\partial}{\partial t} M[\varphi]$ also satisfies the wave equation. Next, from (1.48), for $t > 0$ we get

$$\frac{\partial^2}{\partial t^2} M[\varphi](t,x) = \Delta M[\varphi](t,x) = M[\Delta \varphi](t,x).$$

But, since $M[\Delta \varphi](t,x) \to 0$ as $t \to 0$, we have that

$$\frac{\partial v}{\partial t}(t,x) = \frac{\partial^2}{\partial t^2} M[\varphi](t,x) \longrightarrow 0 \qquad (t \to 0).$$

So, from the above, $v(t, x)$ satisfies the wave equation and the initial conditions

$$v(t, x) \longrightarrow \varphi(x), \qquad \frac{\partial v}{\partial t}(t, x) \longrightarrow 0 \qquad (t \to 0)$$

are also satisfied. Hence, we have shown that (1.41) is the required solution.

Next, we turn to the case $f \in C^2([0, \infty) \times \mathbb{R}^3)$ and determine an expression for the solution of

$$\begin{cases} \Box u = f & \text{in } (0, \infty) \times \mathbb{R}^3, \\ u(0, x) = 0, & \dfrac{\partial u}{\partial t}(0, x) = 0. \end{cases}$$

To accomplish this, we apply Duhamel's principle; then from the work at hand, an expression for the solution follows immediately. We explain this in a little more detail.

To begin with, we set

$$(1.49) \qquad u(t, x) = \int_0^t M[f(s, \cdot)](t - s, x)ds.$$

Then, from our previous results, we have $u \in C^2([0, \infty) \times \mathbb{R}^3)$. Also,

$$\Delta u(t, x) = \int_0^t \Delta(M[f(s, \cdot)])(t - s, x)ds.$$

So, by partially differentiating with respect to t, we see that
(1.50)
$$\frac{\partial u}{\partial t}(t, x) = M[f(t, \cdot)](t - t, x) + \int_0^t \frac{\partial}{\partial t}(M[f(s, \cdot)])(t - s, x)ds$$
$$= \int_0^t \frac{\partial}{\partial t}(M[f(s, \cdot)])(t - s, x)ds.$$

By partially differentiating with respect to t once more, we get

$$\frac{\partial^2 u}{\partial t^2}(t, x) = \left(\frac{\partial}{\partial t} M[f(s, \cdot)] \right)(t - s, x) \Big|_{s=t}$$
$$+ \int_0^t \frac{\partial^2}{\partial t^2}(M[f(s, \cdot)])(t - s, x)ds.$$

Now, if we use $\left(\dfrac{\partial}{\partial t} M[f(s, \cdot)] \right)(0, x) = f(s, x)$, we obtain that

$$(1.51) \quad \Box u(t, x) = f(t, x) + \int_0^t \Box(M[f(s, \cdot)])(t - s, x)dx = f(t, x),$$

and hence $u(t, x)$ satisfies the equation.

Now we take a quick look at the initial conditions. By letting $t \to 0$ in (1.49) and (1.50), it follows that both right-hand sides converge to zero, and so the initial conditions are satisfied.

The 2-dimensional case.

Assume that $\psi \in C^2(\mathbb{R}^2)$, and let $x = (x_1, x_2)$ be a point in $\mathbb{R}^2$. Further, let $p = (x, z) = (x_1, x_2, z)$ be a point in $\mathbb{R}^3$. Continuing our definitions, we define $\widetilde{\psi}$ by

$$\widetilde{\psi}(x, z) = \psi(x);$$

then $\widetilde{\psi} \in C^2(\mathbb{R}^3)$.

Now, set

$$\widetilde{u}(t, p) = M[\widetilde{\psi}](t, p).$$

We have already shown that the following holds:

$$\frac{\partial}{\partial z} \widetilde{u}(t, p) = M\left[\frac{\partial \widetilde{\psi}}{\partial z} \right](t, p), \qquad \forall \, (t, p) \in (0, \infty) \times \mathbb{R}^3.$$

Since $\dfrac{\partial \widetilde{\psi}}{\partial z} \equiv 0$, the function $\widetilde{u}(t, p)$ is independent of z.

On the other hand,

$$\left(\frac{\partial^2}{\partial t^2} - \sum_{j=1}^{2} \frac{\partial^2}{\partial x_j^2} \right) \widetilde{u}(t, p) = \left(\frac{\partial^2 \widetilde{u}}{\partial t^2} - \sum_{j=1}^{2} \frac{\partial^2 \widetilde{u}}{\partial x_j^2} - \frac{\partial^2 \widetilde{u}}{\partial z^2} \right)(t, p) = 0,$$

and so

$$\widetilde{u}(t, p) \longrightarrow 0, \quad \frac{\partial \widetilde{u}}{\partial t}(t, p) \longrightarrow \widetilde{\psi}(p) = \psi(x) \qquad (t \to 0).$$

If we define $u(t, x) \in C^2((0, \infty) \times \mathbb{R}^2)$ by

$$u(t, x) = \widetilde{u}(t, x, 0),$$

then from the above it follows that u satisfies

$$\begin{cases} \left(\dfrac{\partial^2}{\partial t^2} - \sum_{j=1}^{2} \dfrac{\partial^2}{\partial x_j^2} \right) u = 0, \\[2mm] u(t, x) \longrightarrow 0, \quad \dfrac{\partial u}{\partial t}(t, x) \longrightarrow \psi(x) \qquad (t \to 0). \end{cases}$$

Now, we write

$$u(t, x) = M[\widetilde{\psi}](t, x, 0) = \frac{1}{4\pi t} \int_{|(y,z)-(x,0)|=t} \widetilde{\psi}(y, z)\, dS_{y,z}.$$

We can express the part of the sphere of radius t and center at $(x, 0)$ with $z > 0$ as

$$\{(x_1 + s_1, x_2 + s_2, \sqrt{t^2 - s_1^2 - s_2^2}\,); s_1^2 + s_2^2 < t^2\}.$$

In addition, if we set I^+ to be the integral of the upper half sphere, and note that $dS_{y,z} = (t^2 - s_1^2 - s_2^2)^{-1/2} ds_1 ds_2$, then we have that

$$I^+ = \frac{1}{4\pi t} \int_{s_1^2 + s_2^2 < t^2} \frac{\psi(x_1 + s_1, x_2 + s_2)}{\sqrt{t^2 - s_1^2 - s_2^2}}\, ds_1 ds_2;$$

while for I^-, the integral of the lower half sphere, since ψ is independent of z, it turns out to be the same as I^+. So, we can express $I^+ + I^-$ as the integral

$$M_2[\psi](t, x) = \frac{1}{2\pi t} \int_{|y-x|<t} \frac{\psi(y)}{\sqrt{t^2 - |x - y|^2}}\, dy,$$

which is precisely the formula we want.

With regard to the first term in (1.42), since

$$\frac{\partial}{\partial t} M[\widetilde{\psi}](t, x, 0) = \frac{\partial}{\partial t} M_2[\psi](t, x),$$

the required property follows immediately.

Now we take the opportunity to state what we know from Theorem 1.2. For the $n = 3$ case, if we assume that $\operatorname{supp} \varphi \cup \operatorname{supp} \psi \subset \{x; |x - x_0| < \varepsilon\}$ $(\varepsilon > 0)$, then from (1.41) we see that

$$u(t, x) \neq 0 \implies vt - \varepsilon \leqslant |x - x_0| \leqslant vt + \varepsilon.$$

Since $\varepsilon > 0$ can be chosen arbitrarily, we can say that, "For time $t > 0$, the initial data at point x_0 only effects the sphere $\{x; |x - x_0| =$

vt}. That is to say, the effect is propagated with an exact speed of
v."

On the other hand, for the $n = 2$ case, the effect of the initial
data at the point x_0 is in the ball $\{x; |x - x_0| \leqslant vt\}$, and even after
the passage of some time this effect remains at points close to x_0.

As in the $n = 3$ case, if the effect of the initial data at x_0 at time
$t = 0$ is limited to the cone $t^2 = |x - x_0|^2$ in the space (t, x), we say
that *Huygens' principle* holds.

Chapter summary.

1.1 The vibration of a string, the oscillation of a spring, the
propagation of sound, the propagation of electromagnetic waves, the
propagation due to an earthquake, *etc.*; these wave phenomena if
written down as partial differential equations all have basically the
same type of equation.

1.2 The equations of the wave phenomena considered in §1.1
are part of a class of partial differential equations called second-order
hyperbolic equations.

1.3 It is possible to write the solution of the initial value problem for the d'Alembert operator. Using this solution, it is a straightforward matter to investigate the properties of the solution.

Exercises

1. (1) Show that there exist 2 functions f and g, defined on $\mathbb{R}$, such
that the solution of the 1-dimensional wave equation,

$$\Box u(t, x) = \frac{\partial^2 u}{\partial t^2}(t, x) - \frac{\partial^2 u}{\partial x^2}(t, x) = 0,$$

can be written as

$$u(t, x) = f(x - t) + g(x + t).$$

(2) Show that the solution of the initial value problem

$$\begin{cases} \Box u(t, x) = 0 & ((t, x) \in (0, \infty) \times \mathbb{R}), \\ u(0, x) = \varphi(x), & u_t(0, x) = \psi(x), \end{cases}$$

is given by

$$u(t, x) = \frac{1}{2}\left\{\varphi(x - t) + \varphi(x + t) + \int_{x-t}^{x+t} \psi(s)ds\right\}.$$

(3) Show that there exists a function f, defined on $\mathbb{R}$, such that the solution of the boundary value problem

$$\begin{cases} \Box u(t,x) = 0 & ((t,x) \in (-\infty,\infty) \times (0,\infty)), \\ u(t,0) = 0, \end{cases}$$

can be expressed as

$$u(t,x) = f(x+t) - f(x-t).$$

(4) Find an expression for the solution of the initial boundary value problem

$$\begin{cases} \Box u(t,x) = 0 & ((t,x) \in (0,\infty) \times (0,\infty)), \\ u(t,0) = 0, \\ u(0,x) = \varphi(x), \quad u_t(0,x) = \psi(x) & (x \in (0,\infty)), \end{cases}$$

with the assumption that $\varphi(0) = \psi(0) = 0$.

2. Suppose a uniform string is suspended from the ceiling. Set up the equation of motion that traverses this string.

3. Suppose that ω is an element of $\mathbb{R}^3$ such that $|\omega| = 1$, further suppose that $h(l)$ is a real-valued function with continuous derivatives of second order that is defined on $\mathbb{R}$. In the particular case of no external force, show that

$$\boldsymbol{u}^P(t,x) = \omega h(x \cdot \omega - \sqrt{(\alpha+\beta)/\rho}\, t)$$

is a solution of the elastic wave (1.31).
    Further, if we suppose that $\eta \in \mathbb{R}^3$ and ω are orthogonal, show that in the specific case of $\boldsymbol{f} \equiv 0$,

$$\boldsymbol{u}^S(t,x) = \eta h(x \cdot \omega - \sqrt{\alpha/\rho}\, t)$$

is a solution of (1.31). (Note, the speeds of propagation of $\boldsymbol{u}^P$ and $\boldsymbol{u}^S$ are different.)

4. Consider the following initial value problem:

$$\begin{cases} \dfrac{\partial^2 u}{\partial t^2} - a_{11}\dfrac{\partial^2 u}{\partial x_1^2} - a_{22}\dfrac{\partial^2 u}{\partial x_2^2} - a_{33}\dfrac{\partial^2 u}{\partial x_3^2} = 0 & \text{in } (0,\infty) \times \mathbb{R}^3, \\ u(0,x) = 0, \quad u_t(0,x) = \varphi(x). \end{cases}$$

With the proviso that the coefficients are constants that satisfy $a_{11} \geqslant a_{22} \geqslant a_{33} > 0$ and $\varphi \in C^2(\mathbb{R}^3)$:
(1) Determine an expression for the solution.

(2) Suppose $\operatorname{supp}\varphi \subset \{x; |x| \leqslant \varepsilon\}$. For $t > 0$, determine an estimate for $\operatorname{supp} u(t, \cdot)$.

5. Consider Maxwell's equations (1.20)–(1.23) with $\varepsilon_0 = \mu_0 = 1$. Suppose that $\boldsymbol{j}$ and q satisfy $\operatorname{div}\boldsymbol{j} = \dfrac{\partial q}{\partial t}$. Further, suppose that the initial values $\boldsymbol{E}_0(x)$ and $\boldsymbol{B}_0(x)$ satisfy, respectively,

$$\operatorname{div}\boldsymbol{E}_0(x) = q(0, x) \quad \text{and} \quad \operatorname{div}\boldsymbol{B}_0(x) = 0.$$

Finally, let us suppose that $\boldsymbol{B}(t, x)$ is the solution of the initial value problem

$$\begin{cases} \Box \boldsymbol{B}(t, x) = -\operatorname{rot}\boldsymbol{j}, \\ \boldsymbol{B}(0, x) = \boldsymbol{B}_0(x), \quad \dfrac{\partial \boldsymbol{B}}{\partial t}(0, x) = -\operatorname{rot}\boldsymbol{E}_0(x). \end{cases}$$

If we define $\boldsymbol{E}(t, x)$ by

$$\boldsymbol{E}(t, x) = \int_0^t \{\operatorname{rot}\boldsymbol{B}(s, x) + \boldsymbol{j}(s, x)\}ds + \boldsymbol{E}_0(x),$$

show that $\{\boldsymbol{E}(t, x), \boldsymbol{B}(t, x)\}$ is a solution of Maxwell's equations with the initial value $\{\boldsymbol{E}_0(x), \boldsymbol{B}_0(x)\}$.

CHAPTER 2

The Existence of a Solution for a Hyperbolic Equation and its Properties

In this chapter, we study the basic properties of a single linear second-order hyperbolic equation. But first we backtrack for a few moments, and once again consider the wave phenomena. By applying, for instance, an external force to some part of whatever we are considering, we can cause a disturbance. With time, this disturbance is passed onto the surrounding parts. One particular characteristic of wave phenomena is that the speed with which the disturbance propagates itself is finite. This is in stark contrast to how heat propagates. In fact, we shall show that finiteness of the speed of propagation is a property of a second-order hyperbolic equation. Having accomplished this, we will then show that a solution for the initial boundary value problem does exist. In this proof of existence, we will rely heavily on the so-called *a priori* estimate to establish an estimate for the solution.

2.1 Finite propagation speed, domains of dependence and influence

We are familiar with the fact that sound travels at 340m/sec and that a radio wave circles the earth 7 and a half times per 1 second. Regardless of whether it is sound being propagated or an electromagnetic wave, the conclusion we can draw from the above is that in both cases the speed of propagation is constant.

As we explained in the first chapter, both the propagation of sound and electromagnetic waves can be described by means of wave equations and both of these waves have a constant speed of propagation. Now, the wave equation is a typical example of a hyperbolic equation. In particular, one characteristic of the solution of wave equations is that the speed of propagation is finite, and, as expected, this is a common feature of the solutions of hyperbolic equations.

In this chapter, we would like to consider the following second-order hyperbolic operator:

(2.1)
$$P = \frac{\partial^2}{\partial t^2} + 2\sum_{j=1}^{n} h_j(t,x)\frac{\partial^2}{\partial t \partial x_j} - \sum_{j,l=1}^{n} a_{jl}(t,x)\frac{\partial^2}{\partial x_j \partial x_l}$$
$$+ \sum_{j=1}^{n} a_j(t,x)\frac{\partial}{\partial x_j} + h_0(t,x)\frac{\partial}{\partial t} + a_0(t,x).$$

The reader might note that the above is just equation (1.33) with $N = 1$. In (2.1), all the coefficients belong to $\mathcal{B}^\infty(\mathbb{R} \times \mathbb{R}^n)$, and the coefficients of the principal parts, namely h_j $(j = 1, 2, \dots, n)$ and a_{jl} $(j, l = 1, 2, \dots, n)$, are all real-valued functions. Further, $(a_{jl})_{j,l=1,2,\dots,n}$ are what are termed *elliptic*, by this we mean that

(2.2)
$$\sum_{j,l=1}^{n} a_{jl}(t,x)\xi_j\xi_l \geqslant c_0 \sum_{j=1}^{n} \xi_j^2$$

for all (t,x) and all $\xi = (\xi_1, \dots, \xi_n) \in \mathbb{R}^n$, and where c_0 is a positive constant.

The characteristic polynomial, p_0, of P is

(2.3)
$$p_0(t,x,\lambda,\xi) = \lambda^2 + 2\sum_{j=1}^{n} h_j\xi_j\lambda - \sum_{j,l=1}^{n} a_{jl}\xi_j\xi_l.$$

The characteristic roots, λ^+ and λ^-, of p_0 are

(2.4)
$$\lambda^{\pm}(t,x,\xi) = \sum_{j=1}^{n} h_j\xi_j \pm \left\{ \left(\sum_{j=1}^{n} h_j\xi_j\right)^2 + \sum_{j,l=1}^{n} a_{jl}\xi_j\xi_l \right\}^{1/2}.$$

However, since (a_{jl}) are elliptic, it follows that $\lambda^{\pm}$ are real-valued and satisfy
$$|\lambda^+ - \lambda^-| \geqslant 2\sqrt{c_0}|\xi|.$$

Hence, P is a normal hyperbolic operator.

To provide clarity in our further discussions, we introduce the following notation:

(2.5)
$$A = \sum_{j,l=1}^{n} a_{jl}(t,x)\frac{\partial^2}{\partial x_j \partial x_l} - \sum_{j=1}^{n} a_j(t,x)\frac{\partial}{\partial x_j} - a_0(t,x)$$

and

$$(2.6) \qquad H = 2\sum_{j=1}^{n} h_j(t,x)\frac{\partial}{\partial x_j} + h_0.$$

In terms of (2.5) and (2.6), we can write the operator P as

$$(2.7) \qquad Pu = \frac{\partial^2 u}{\partial t^2} + H\frac{\partial u}{\partial t} - Au.$$

Now, we suppose that Ω is a domain in $\mathbb{R}^n$ with a smooth boundary Γ. Our intention is to consider the initial boundary value problem for P relative to this domain. To be more explicit, the problem is one of finding a $u(t,x)$ that satisfies

$$(2.8) \qquad \begin{cases} Pu(t,x) = f(t,x) & ((t,x) \in (0,\infty) \times \Omega), \\ Bu(t,x) = 0 & ((t,x) \in (0,\infty) \times \Gamma), \\ u(0,x) = u_0(x), \quad \dfrac{\partial u}{\partial t}(0,x) = u_1(x) & (x \in \Omega), \end{cases}$$

where $f(t,x)$ is a given function on $(0,\infty) \times \Omega$ and $u_0(x), u_1(x)$ are given functions on Ω. In the above, the boundary operator is considered to be either

$$(2.9) \qquad B_D u = u$$

or

$$(2.10) \qquad B_N u = \sum_{j,l=1}^{n} a_{jl}(t,x)\nu_l(x)\frac{\partial u}{\partial x_j} - \sigma_0(t,x)\frac{\partial u}{\partial t} + \sigma_1(t,x)u,$$

where $\nu(x) = (\nu_1(x), \nu_2(x), \ldots, \nu_n(x))$ denotes the outward unit normal vector for Ω at $x \in \Gamma$, and σ_0, σ_1 are functions in $\mathcal{B}^\infty(\mathbb{R} \times \Gamma)$. In addition, we suppose that σ_0 is a real-valued function that satisfies the following:

$$(2.11) \qquad \sigma_0(t,x) \leqslant \sum_{j=1}^{n} h_j(t,x)\nu_j(x) \qquad ((t,x) \in \mathbb{R} \times \Gamma).$$

If we use the notation

$$(2.12) \qquad \frac{\partial}{\partial \nu_A} = \sum_{j,l=1}^{n} a_{jl}(t,x)\nu_l(x)\frac{\partial}{\partial x_j}$$

in the defining equation, (2.10), for B_N, then we can write

$$(2.13) \qquad B_N = \frac{\partial}{\partial \nu_A} - \sigma_0 \frac{\partial}{\partial t} + \sigma_1.$$

When the boundary operator, B_D, is given by (2.9), the boundary condition is called *Dirichlet's boundary condition*, while when the boundary operator is given by (2.10) the boundary condition is called the *third boundary condition* or the *Robin condition*. In this book, we will focus our attention solely on the third boundary condition. The reason for this is that Dirichlet's boundary condition can be dealt with relatively easily, and so does not really warrant significant consideration.

In order not to overcomplicate matters, we restrict ourselves to either the domain Ω given by $\mathbb{R}_+^n = \{(x_1, x_2, \ldots, x_n); x_j \in \mathbb{R} \ (j = 1, 2, \ldots, n-1), x_n > 0\}$ or to the case when Γ is a smooth closed surface. Also, by means of a change of the unknown function, we should note that for the boundary operator, B_N, given by (2.10), we need only investigate the specific case of $\sigma_1 \equiv 0$. We explain why this is so in the case of $\Omega = \mathbb{R}_+^n$.

We begin by taking $\chi(l) \in C^\infty([0, \infty))$ to be such that $\operatorname{supp} \chi \subset [0, 1)$, $\chi(l) = 1$ for $l \in [0, 1/2]$, and then set

$$u(t, x) = \exp\left(x_n \chi(x_n) \frac{\sigma_1(t, x')}{a_{nn}(t, x')}\right) w(t, x)$$

$$= \alpha(t, x) w(t, x).$$

From the fact that we have $\alpha \equiv 1$ on Γ, we see that

$$B_N u|_{x_n=0} = B_N w|_{x_n=0} - \sigma_1(t, x') w|_{x_n=0}.$$

Hence, if $B_N u|_{x_n=0} = 0$, then it follows that w satisfies the boundary condition with $\sigma_1 \equiv 0$. On the other hand, the principal part of the equation that w must satisfy on $(0, \infty) \times \Omega$ is the same as that for P.

So, to reiterate, from this point on we will restrict our attention to the problem with $\sigma_1 \equiv 0$. Also, without any further reminder, we will take the boundary condition to be the third boundary condition, and we will denote B_N by B.

(a) Energy estimate of the solution.

If a solution exists for the initial boundary value problem (2.8) for the hyperbolic operator, P, it is possible to find an estimate that this solution necessarily satisfies. As a consequence of this estimate, the phenomena governed by (2.8) can be shown to have a finite speed of propogation.

As a first step in this direction, we note that the following equations hold; the direct computations are left to the reader.

$$2\operatorname{Re}\frac{\partial^2 u}{\partial t^2}\frac{\overline{\partial u}}{\partial t} = \frac{\partial}{\partial t}\left(\left|\frac{\partial u}{\partial t}\right|^2\right),$$

$$2\operatorname{Re}h_j\frac{\partial^2 u}{\partial x_j \partial t}\frac{\overline{\partial u}}{\partial t} = \frac{\partial}{\partial x_j}\left(h_j\left|\frac{\partial u}{\partial t}\right|^2\right) - \frac{\partial h_j}{\partial x_j}\left|\frac{\partial u}{\partial t}\right|^2,$$

$$-2\operatorname{Re}a_{jj}\frac{\partial^2 u}{\partial x_j^2}\frac{\overline{\partial u}}{\partial t}$$

$$= -2\operatorname{Re}\frac{\partial}{\partial x_j}\left(a_{jj}\frac{\partial u}{\partial x_j}\frac{\overline{\partial u}}{\partial t}\right) + \operatorname{Re}\frac{\partial}{\partial t}\left(a_{jj}\left|\frac{\partial u}{\partial x_j}\right|^2\right)$$

$$+ 2\operatorname{Re}\left\{\frac{\partial a_{jj}}{\partial x_j}\frac{\partial u}{\partial x_j}\frac{\overline{\partial u}}{\partial t} - \frac{\partial a_{jj}}{\partial t}\left|\frac{\partial u}{\partial x_j}\right|^2\right\},$$

and for $j \neq l$

$$-2\operatorname{Re}a_{jl}\frac{\partial^2 u}{\partial x_j \partial x_l}\frac{\overline{\partial u}}{\partial t}$$

$$= -\operatorname{Re}\frac{\partial}{\partial x_j}\left(a_{jl}\frac{\partial u}{\partial x_l}\frac{\overline{\partial u}}{\partial t}\right) - \operatorname{Re}\frac{\partial}{\partial x_l}\left(a_{jl}\frac{\partial u}{\partial x_j}\frac{\overline{\partial u}}{\partial t}\right)$$

$$+ \operatorname{Re}\frac{\partial}{\partial t}\left(a_{jl}\frac{\partial u}{\partial x_j}\frac{\overline{\partial u}}{\partial x_l}\right)$$

$$+ \operatorname{Re}\left(\frac{\partial a_{jl}}{\partial x_j}\frac{\partial u}{\partial x_l}\frac{\overline{\partial u}}{\partial t} + \frac{\partial a_{jl}}{\partial x_l}\frac{\partial u}{\partial x_j}\frac{\overline{\partial u}}{\partial t} - \frac{\partial a_{jl}}{\partial t}\frac{\partial u}{\partial x_j}\frac{\overline{\partial u}}{\partial x_l}\right).$$

Then, using the above formulae, we obtain

$$(2.14) \quad 2\operatorname{Re}Pu\frac{\overline{\partial u}}{\partial t} = \frac{\partial}{\partial t}e(u;t,x) + \sum_{j=1}^{n}\frac{\partial}{\partial x_j}X_j(u;t,x) + Z(u;t,x),$$

where

$$(2.15) \quad e(u;t,x) = \left|\frac{\partial u}{\partial t}(t,x)\right|^2 + \sum_{j,l=1}^{n}a_{jl}(t,x)\frac{\partial u}{\partial x_j}(t,x)\frac{\overline{\partial u}}{\partial x_l}(t,x)$$

and

$$
X_j(u; t, x) = 2h_j(t, x) \left| \frac{\partial u}{\partial t}(t, x) \right|^2
$$

(2.16)

$$
- 2\mathrm{Re} \sum_{l=1}^{n} a_{jl}(t, x) \frac{\partial u}{\partial x_l}(t, x) \overline{\frac{\partial u}{\partial t}(t, x)}
$$

and, finally, Z is the collection of terms that are not included in $\frac{\partial}{\partial t} e$ and $\sum \frac{\partial}{\partial x_j} X_j$ in the terms expressed in the right-hand side of the equation dealt with previously. From the latter, we have that there exists a $C > 0$ such that the following estimate holds:

(2.17)

$$
|Z(u; t, x)| \leqslant C \left(\left| \frac{\partial u}{\partial t}(t, x) \right|^2 + \sum_{j=1}^{n} \left| \frac{\partial u}{\partial x_j}(t, x) \right|^2 + |u(t, x)|^2 \right).
$$

If we make use of (2.2) in the above e, we see that

(2.18)

$$
e(u; t, x) \geqslant c \left\{ \left| \frac{\partial u}{\partial t}(t, x) \right|^2 + \sum_{j=1}^{n} \left| \frac{\partial u}{\partial x_j}(t, x) \right|^2 \right\},
$$

where $c = \min(c_0, 1) > 0$.

Now set

(2.19)

$$
v_{\max} = \sup |\lambda_+(t, x, \xi)|,
$$

where the supremum is taken over all $(t, x) \in \mathbb{R} \times \Omega$ and $|\xi| = 1$.

Now, since $\lambda^+(t, x, -\xi) = -\lambda^-(t, x, \xi)$ it follows that $\sup |\lambda^+| = \sup |\lambda^-|$, and since $\sum_{j=1}^{n} h_j(-\xi_j) > 0$ if $\sum h_j \xi_j < 0$, we get that

(2.20)

$$
v_{\max} \geqslant \left| \sum_{j=1}^{n} h_j \xi_j \right| + \left\{ \left(\sum_{j=1}^{n} h_j \xi_j \right)^2 + \sum_{j,l=1}^{n} a_{jl} \xi_j \xi_l \right\}^{1/2}
$$

holds for all $|\xi| = 1$.

For $(t_0, x_0) \in \mathbb{R} \times \overline{\Omega}$, let

$$\Lambda(t_0, x_0) = \{(t, x); |x - x_0| \leqslant v_{\max}(t_0 - t)\},$$
$$\widetilde{\Lambda}(t_0, x_0) = \Lambda(t_0, x_0) \cap (\mathbb{R} \times \Omega).$$

Now, for chosen fixed $t_0 > t_2 > t_1$ and for $\tau \in [t_1, t_2]$, we set

$$D(\tau) = \widetilde{\Lambda}(t_0, x_0) \cap \{t = \tau\}$$
$$= \{(\tau, x); |x - x_0| \leqslant v_{\max}(t_0 - \tau), x \in \Omega\}.$$

Also, we set

$$V(t_1, t_2) = \Lambda(t_0, x_0) \cap ([t_1, t_2] \times \Omega) = \bigcup_{\tau \in [t_1, t_2]} D(\tau),$$
$$S(t_1, t_2) = \partial \Lambda(t_0, x_0) \cap ([t_1, t_2] \times \Omega),$$
$$S_b(t_1, t_2) = \Lambda(t_0, x_0) \cap ([t_1, t_2] \times \Gamma).$$

So, $\partial V = D(t_1) \cup D(t_2) \cup S \cup S_b$.

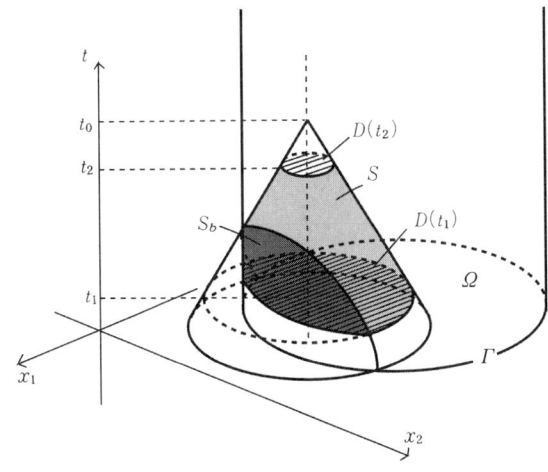

FIGURE 2.1

Next we integrate $2\mathrm{Re}\, Pu\overline{\dfrac{\partial u}{\partial t}}$ over V. By applying the divergence theorem, we obtain

(2.21)
$$
\begin{aligned}
\int_V 2\mathrm{Re}\, Pu\overline{\frac{\partial u}{\partial t}}dtdx = & \int_{D(t_2)} e(u;t_2,x)dx - \int_{D(t_1)} e(u;t_1,x)dx \\
& + \int_{S_b} \sum_{j=1}^n X_j(u;t,x)\nu_j dS \\
& + \int_S \left\{ e(u;t,x)\eta + \sum_{j=1}^n X_j(u;t,x)\mu_j \right\} dS \\
& + \int_V Z(u;t,x)dtdx ,
\end{aligned}
$$

where dS denotes the surface element of the surface S or the surface S_b, the vector $(\eta,\mu_1,\mu_2,\ldots,\mu_n) \in \mathbb{R}^{n+1}$ is the outward unit normal vector of S with respect to V at $(t,x) \in S$, and $\nu = (\nu_1,\nu_2,\ldots,\nu_n) \in \mathbb{R}^n$ is the outward unit normal vector of Γ at $x \in \Gamma$.

Since

$$
\sum_{j=1}^n X_j \nu_j = -2\mathrm{Re}\left(\sum_{j,l=1}^n a_{jl}\nu_j \frac{\partial u}{\partial x_j} - \sum_{j=1}^n h_j \nu_j \frac{\partial u}{\partial t} \right) \overline{\frac{\partial u}{\partial t}}
$$

by using (2.10), we have that

$$
\sum_{j=1}^n X_j(u;t,x)\nu_j = -2\mathrm{Re}\, Bu \cdot \overline{\frac{\partial u}{\partial t}} + \left(\sum_{j=1}^n h_j \nu_j - \sigma_0 \right) \left| \frac{\partial u}{\partial t} \right|^2 .
$$

However, as we noted previously, $\sigma_1 \equiv 0$, and so if we use (2.11) with this, we see that

(2.22)
$$
\int_{S_b} \sum_{j=1}^n X_j(u;t,x)\nu_j dS \geqslant -2\mathrm{Re} \int_{S_b} Bu\overline{\frac{\partial u}{\partial t}}dS.
$$

Next, we need to determine an estimate of the integral over S. Since we know that $(\eta,\mu_1,\ldots,\mu_n)$ is a normal vector of $\Lambda(t_0,x_0)$, we have

(2.23)
$$
\eta = v_{\max}\left(\sum_{j=1}^n \mu_j^2 \right)^{1/2}.
$$

By using the fact that (a_{jl}) satisfies (2.2), we have that for arbitrary $(\xi_1, \xi_2, \ldots, \xi_n)$ and $(\psi_1, \psi_2, \ldots, \psi_n) \in \mathbb{C}^n$ the following can be shown to be true:

$$(2.24) \quad \left| \sum_{j,l=1}^{n} a_{jl} \xi_j \overline{\psi_l} \right| \leqslant \left(\sum_{j,l=1}^{n} a_{jl} \xi_j \overline{\xi_l} \right)^{1/2} \left(\sum_{j,l=1}^{n} a_{jl} \psi_j \overline{\psi_l} \right)^{1/2}.$$

This inequality can be derived as follows. For an arbitrary $\lambda \in \mathbb{C}$, since

$$\sum_{j,l} a_{jl}(\xi_j + \lambda \psi_j)\overline{(\xi_l + \lambda \psi_l)} \geqslant 0,$$

we have that

$$|\lambda|^2 \left(\sum_{j,l=1}^{n} a_{jl} \psi_j \overline{\psi_l} \right) - \left(\overline{\lambda} \sum_{j,l=1}^{n} a_{jl} \xi_j \overline{\psi_l} + \lambda \sum_{j,l=1}^{n} a_{jl} \psi_l \overline{\xi_j} \right)$$
$$+ \sum_{j,l=1}^{n} a_{jl} \xi_j \overline{\xi_l} \geqslant 0.$$

Suppose that $\sum a_{jl} \psi_j \overline{\psi_l} > 0$ and take

$$\lambda = \left(\sum a_{jl} \overline{\psi_l} \xi_j \right)\left(\sum a_{jl} \psi_l \overline{\psi_j} \right)^{-1}.$$

Then

$$-\left| \sum a_{jl} \overline{\psi_l} \xi_j \right|^2 \left(\sum a_{jl} \psi_j \overline{\psi_l} \right)^{-1} + \sum a_{jl} \xi_j \overline{\xi_l} \geqslant 0,$$

which in turn gives (2.24).

Turning our attention to the given forms of e and X_j, we see that the integrand over S of the right-hand side of (2.21) becomes

$$(2.25)$$
$$\eta \left(\left| \frac{\partial u}{\partial t} \right|^2 + \sum_{j,l=1}^{n} a_{jl} \frac{\partial u}{\partial x_j} \overline{\frac{\partial u}{\partial x_l}} \right) + 2 \sum_{j=1}^{n} h_j \mu_j \left| \frac{\partial u}{\partial t} \right|^2$$
$$- \operatorname{Re} \sum_{j,l=1}^{n} a_{jl} \frac{\partial u}{\partial x_l} \mu_j \overline{\frac{\partial u}{\partial t}}.$$

For the final term in (2.25), by letting $\xi_j = \dfrac{\partial u}{\partial x_j}$, $\psi_l = \mu_l \dfrac{\partial u}{\partial t}$ and applying (2.24), we obtain the following estimate:

(2.26)

$$\left| \mathrm{Re} \sum_{j,l=1}^{n} a_{jl} \frac{\partial u}{\partial x_j} \mu_l \overline{\frac{\partial u}{\partial t}} \right|$$

$$\leqslant \left(\sum_{j,l=1}^{n} a_{jl} \frac{\partial u}{\partial x_j} \overline{\frac{\partial u}{\partial x_l}} \right)^{1/2} \left(\sum_{j,l=1}^{n} a_{jl} \mu_j \mu_l \right)^{1/2} \left| \frac{\partial u}{\partial t} \right|.$$

In the above, we set

$$L = \sum_{j,l=1}^{n} a_{jl} \frac{\partial u}{\partial x_j} \overline{\frac{\partial u}{\partial x_l}}, \quad M = \sum_{j,l=1}^{n} a_{jl} \mu_j \mu_l, \quad N = \sum_{j=1}^{n} h_j \mu_j.$$

Since $\eta + 2N > 0$ is a consequence of (2.20) and (2.23), and then by using (2.26), we can estimate that

(2.27) the above (2.25) $\geqslant (\eta + 2N) \left| \dfrac{\partial u}{\partial t} \right|^2 - 2L^{1/2} M^{1/2} \left| \dfrac{\partial u}{\partial t} \right| + \eta L.$

Regarding the right-hand side of the above as a quadratic expression in $\left| \dfrac{\partial u}{\partial t} \right|$, we see that if

(2.28) $(L^{1/2} M^{1/2})^2 \leqslant (\eta + 2N)\eta L$

is satisfied, then the right-hand side of (2.27) is non-negative for all values of $\left| \dfrac{\partial u}{\partial t} \right|$ and therefore (2.25) is also non-negative. Now, (2.28) can be written as

(2.29) $\eta^2 + 2N\eta - M \geqslant 0.$

But, the left-hand side is the characteristic equation $p_0(t, x, \eta, \mu)$ of P. Therefore, if we combine definition (2.19) of $v_{\max}$ with the relation (2.23) for η and μ, we see that (2.29) does in fact hold. So what the above shows is that the integrand over S of the right-hand side of (2.21) is non-negative.

Now, by using (2.22) and the fact that the integral on S is non-negative in (2.21), the following estimate holds:

(2.30)
$$\int_{D(t_2)} e(u; t_2, x)dx \leqslant \int_{D(t_1)} e(u; t_1, x)dx - \int_V Z(u; t, x)dtdx$$
$$+ 2\mathrm{Re} \int_{S_b} Bu\frac{\overline{\partial u}}{\partial t}dS + 2\mathrm{Re} \int_V Pu\frac{\overline{\partial u}}{\partial t}dtdx.$$

We would like to note that for $t \geqslant t_1$ the following is true:

(2.31)
$$|u(t, x)|^2 = |u(t_1, x)|^2 + \int_{t_1}^t \frac{\partial}{\partial s}|u(s, x)|^2 ds$$
$$\leqslant |u(t_1, x)|^2 + \int_{t_1}^t \left|\frac{\partial u}{\partial s}(s, x)\right|^2 ds + \int_{t_1}^t |u(s, x)|^2 ds.$$

Since $V = \bigcup_{t \in [t_1, t_2]} D(t)$, we can write

$$\int_V Z(u; t, x)dtdx = \int_{t_1}^{t_2} dt \int_{D(t)} Z(u; t, x)dx.$$

By using (2.17) in the right-hand side, and, furthermore, by making use of (2.18), we see that there exists a constant C such that

(2.32) $$\left|\int_V Z(u; t, x)dtdx\right| \leqslant C \int_{t_1}^{t_2} dt \int_{D(t)} (e(u; t, x) + |u(t, x)|^2)dx.$$

In addition, C is a positive constant that is independent of t_1, t_2 and u.

Similarly, we can obtain the following estimate:

$$\left|2\mathrm{Re} \int_V Pu\frac{\overline{\partial u}}{\partial t}dtdx\right|$$
$$\leqslant \int_V |Pu|^2 dtdx + \int_V \left|\frac{\partial u}{\partial t}(t, x)\right|^2 dtdx$$
$$\leqslant \int_{t_1}^{t_2} dt \int_{D(t)} |Pu(t, x)|^2 dx + \int_{t_1}^{t_2} dt \int_{D(t)} e(u; t, x)dx.$$

Now set

(2.33) $$\tilde{e}(u; t, x) = e(u; t, x) + |u(t, x)|^2.$$

If we substitute (2.32) into (2.30) and combine this with (2.31), then the result is the next proposition.

PROPOSITION 2.1. *Suppose* u *is a function defined in* $\overline{\Lambda(t_0, x_0)}$ *with continuous partial derivatives up to second order. Then for arbitrary* t_1 *and* t_2 *such that* $t_1 < t_2 < t_0$, *the following estimate holds:*

(2.34)

$$\int_{D(t_2)} \widetilde{e}(u; t_2, x)dx$$

$$\leqslant \int_{D(t_1)} \widetilde{e}(u; t_1, x)dx + C \int_{t_1}^{t_2} dt \int_{D(t)} \widetilde{e}(u; t, x)dx$$

$$+ 2\int_{S_b} \left| Bu \frac{\overline{\partial u}}{\partial t} \right| dS + \int_{t_1}^{t_2} dt \int_{D(t)} |(Pu)(t, x)|^2 dx.$$

Next, we consider the following state:

(2.35)

$$\begin{cases} Pu(t, x) = 0, & (t, x) \in V, \\ Bu(t, x) = 0, & (t, x) \in S_b, \\ u(t_1, x) = 0, \quad \dfrac{\partial u}{\partial t}(t_1, x) = 0, & x \in D(t_1). \end{cases}$$

More explicitly, at time t_1, u is at a state of rest in $D(t_1)$; there is no external force applied within the region V, and the boundary conditions are satisfied on S_b.

By using (2.35) in Proposition 2.1, we can see that in the right-hand side of (2.34) all the terms are zero except the second term. In other words, it is easy to see that the following estimate holds:

(2.36) $$\int_{D(t_2)} \widetilde{e}(u; t_2, x)dx \leqslant C \int_{t_1}^{t_2} dt \int_{D(t)} \widetilde{e}(u; t, x)dx.$$

We now need to state and prove the following lemma.

LEMMA 2.2 (Gronwall inequality). *Suppose* $\gamma(t)$ *and* $\rho(t)$ *are non-negative functions defined on the interval* $[0, a]$ *with* $a > 0$, *and, further, suppose* $\gamma(t)$ *is integrable over* $[0, a]$, *and* $\rho(t)$ *is monotonically increasing. In addition, suppose that* $\mu(t)$ *is a real-valued continuous function defined on* $[0, a]$. *If* $0 \leqslant \tau < a$ *and*

(2.37) $$\gamma(t) \leqslant c \int_{\tau}^{t} \gamma(s)ds + \rho(t) + \int_{\tau}^{t} \mu(s)ds \qquad (t \in [\tau, a]),$$

with the proviso that c is a positive constant, then the following is true:

$$(2.38) \qquad \gamma(t) \leqslant e^{c(t-\tau)}\rho(t) + \int_\tau^t e^{c(t-s)}\mu(s)ds \qquad (t \in [\tau, a]).$$

PROOF. Let $G(t) = \int_\tau^t \gamma(s)ds$. Then $G(t)$ is an absolutely continuous function and

$$G'(t) = \gamma(t) \qquad \text{a.e. } t \in [\tau, a].$$

By using $G(t)$ to rewrite (2.37), we see that

$$G'(t) \leqslant c\,G(t) + \rho(t) + \int_\tau^t \mu(s)ds \qquad \text{a.e. } t \in [\tau, a].$$

Since the following is true for t almost everywhere,

$$\frac{d}{dt}(e^{-ct}G(t)) = e^{-ct}\{G'(t) - c\,G(t)\}$$

$$\leqslant e^{-ct}\left\{\rho(t) + \int_\tau^t \mu(s)ds\right\},$$

we can integrate this inequality over the interval $[\tau, t]$ to get

$$e^{-ct}G(t) - G(\tau) \leqslant \int_\tau^t e^{-cs}\rho(s)ds + \int_\tau^t \left(e^{-cs}\int_\tau^s \mu(r)dr\right)ds.$$

By using the fact that $G(\tau) = 0$ and ρ is monotonically increasing, then if the second term of the right-hand side is integrated by parts, we see that

$$G(t) \leqslant \int_\tau^t e^{c(t-s)}\rho(s)ds + \int_\tau^t e^{c(t-s)}\left(\int_\tau^s \mu(r)dr\right)ds$$

$$\leqslant \frac{1}{c}\rho(t)(e^{c(t-\tau)} - 1) - \frac{1}{c}\int_\tau^t \mu(s)ds + \frac{1}{c}\int_\tau^t e^{c(t-s)}\mu(s)ds.$$

If we make use of the above estimate in (2.37), then we obtain

$$\gamma(t) \leqslant c\,G(t) + \rho(t) + \int_\tau^t \mu(s)ds = e^{c(t-\tau)}\rho(t) + \int_\tau^t e^{c(t-s)}\mu(s)ds.$$

$$\square$$

Now, we return our attention to the estimate of $\tilde{e}(u; t_2, x)$. To begin with, we set

$$\gamma(t) = \int_{D(t)} \tilde{e}(u; t, x)dx.$$

Since the t_2 in (2.36) is arbitrary but with $t_1 < t_2 < t_0$, it is easy to see that

$$\gamma(t) \leqslant C \int_{t_1}^{t} \gamma(s)ds \qquad \forall t \in [t_1, t_0].$$

By applying the Gronwall inequality (Lemma 2.2) to this, we get that

$$\gamma(t) = 0 \qquad \forall t \in [t_1, t_0).$$

So, we have

$$\tilde{e}(u; t, x) = 0 \qquad \forall x \in D(t),\ \forall t \in [t_1, t_0).$$

Hence, from the definition of $\tilde{e}$, (2.33), it follows that u is zero in V.

We gather the above together into the following theorem.

THEOREM 2.3. *Suppose that $t_1 < t_0$ and $x_0 \in \overline{\Omega}$. Further, we suppose that $u \in C^2(\mathbb{R} \times \overline{\Omega})$ satisfies*

$$\begin{cases} Pu = 0 & \text{in } \Lambda(t_0, x_0) \cap ([t_1, \infty) \times \Omega), \\ Bu = 0 & \text{on } \Lambda(t_0, x_0) \cap ([t_1, \infty) \times \Gamma), \\ u(t_1, x) = 0, \quad \dfrac{\partial u}{\partial t}(t_1, x) = 0 & \text{for } x \in D(t_1). \end{cases}$$

Then

$$u = 0 \qquad \text{in } \widetilde{\Lambda}(t_0, x_0) \cap \{t \geqslant t_1\}.$$

We note that Theorem 2.3 is a fundamental result concerning the properties of the solution of hyperbolic equations.

(b) The domain of dependence.

We consider the following initial boundary value problem:

(2.39)
$$\begin{cases} Pu = f & \text{in } (0, \infty) \times \Omega, \\ Bu = g & \text{on } (0, \infty) \times \Gamma, \\ u(0, x) = u_0(x), \quad \dfrac{\partial u}{\partial t}(0, x) = u_1(x) & \text{in } \Omega. \end{cases}$$

More explicitly, at initial time $t = 0$ the initial states u_0 and u_1 have been assigned, and we are given f and the condition g on the boundary. As already mentioned in §1.2, the problem is to find a solution that fulfills these conditions.

Now, we select a point $(t_0, x_0) \in (0, \infty) \times \Omega$. It is our intention to examine how the value of the solution u of (2.39) at this point depends on the data (u_0, u_1, f, g). To avoid confusion, we set up the following notation. To emphasize that $D(t)$ depends on (t_0, x_0), we will designate (t_0, x_0) by p_0 and write $D(p_0, t)$. Also, we write $V(p_0, t_1, t_2)$ for V and $S_b(p_0, t_1, t_2)$ for S_b, again to emphasize that they depend on p_0, t_1 and t_2. We can now conclude the following.

THEOREM 2.4. *At the point* $p_0 = (t_0, x_0)$ *the value of the solution* u *of (2.39) is determined only by the values of the initial data* u_0, u_1 *in* $D(p_0, 0)$, *the value of* f *in* $V(p_0, 0, t_0)$ *and the value of* g *in* $S_b(p_0, 0, t_0)$.

PROOF. To start with, we take two sets of data (u_0, u_1, f, g) and (v_0, v_1, h, k) of (2.39), and suppose that, respectively, u and v are the solutions for these two sets of data. Then, we want to show that if these two sets of data satisfy

(2.40)
$$\begin{cases} u_0(x) = v_0(x), \quad u_1(x) = v_1(x), \qquad \forall x \in D(p_0, 0), \\ f(t, x) = h(t, x), \qquad \forall (t, x) \in V(p_0, 0, t_0), \\ g(t, x) = k(t, x), \qquad \forall (t, x) \in S_b(p_0, 0, t_0), \end{cases}$$

we will obtain that

(2.41)
$$u(t_0, x_0) = v(t_0, x_0).$$

With this in mind, we set

$$z(t, x) = u(t, x) - v(t, x).$$

From assumption (2.40), it follows that

$$\begin{cases} Pz = 0 \quad \text{in } V(p_0, 0, t_0), \\ Bz = 0 \quad \text{on } S_b(p_0, 0, t_0), \\ z(0, x) = 0, \quad \dfrac{\partial z}{\partial t}(0, x) = 0 \qquad \text{on } D(p_0, 0). \end{cases}$$

By applying Theorem 2.3 to this, we can deduce that

$$z(t, x) = 0, \qquad \forall (t, x) \in V(p_0, 0, t_0).$$

Since $p_0 \in V(p_0, 0, t_0)$, we have that $z(t_0, x_0) = 0$. In other words, we have shown that (2.41) holds.

$\square$

The above theorem shows that the value of the solution at every point $(t, x) \in (0, \infty) \times \Omega$ is determined solely by its value in some finite domain of the data, and it is not effected by the state outside of this domain. This fact gives rise to the concept which we shall call the domain of dependence for an initial boundary value problem.

DEFINITION 2.5. For the initial boundary value problem (2.39), the set W is the *domain of dependence* of the solution at $(t, x) \in (0, \infty) \times \Omega$, when it possesses the following two properties.

(i) No matter how we choose $\widetilde{W}$, a neighbourhood of W, and how the value of the data outside $\widetilde{W}$ is manipulated, the value of the solution at (t, x) remains completely unchanged.

(ii) There does not exist a set that is smaller than W and possesses property (i).

Let us investigate the domains of dependence for the initial value problem, as discussed in §1.3, of the wave equation in the cases of 2-dimensional space and 3-dimensional space. From the solution of (1.41) and (1.42) the following is easy to deduce.

The 3-dimensional case. The domain of dependence $W(t_0, x_0)$ at $(t_0, x_0) \in (0, \infty) \times \mathbb{R}^3$ is

$$W(t_0, x_0) = \{(t, x) \in [0, \infty) \times \mathbb{R}^3; |x - x_0| = v(t_0 - t)\}.$$

The 2-dimensional case.

$$W(t_0, x_0) = \{(t, x) \in [0, \infty) \times \mathbb{R}^2; |x - x_0| \leqslant v(t_0 - t)\}.$$

For the 3-dimensional case the domain of dependence is only the boundary of a circular cone, while for the 2-dimensional case the domain of dependence also includes the interior of the circular cone. This highlights the fact that even for the same wave equation the domains of dependence change significantly depending on the dimension of the space. So, to determine the exact nature of the domain of dependence is in general a difficult problem. However, for the initial boundary value problem (2.39) at every point $(t_0, x_0) \in (0, \infty) \times \Omega$, Theorem 2.4 implies that

The domain of dependence at $(t_0, x_0) \subset V((t_0, x_0), 0, t_0)$.

(c) The domain of influence.

Let us now look at the mutually complementary concept to the domain of dependence, that is, the domain of influence.

DEFINITION 2.6. For the the initial boundary value problem (2.39), the set Y is a *domain of influence* at $(t, x) \in [0, \infty) \times \overline{\Omega}$ if Y possesses the following two properties.

(i) For arbitrary $\rho \in Y$ and an arbitrary neighbourhood U of (t, x), if we suitably change the value of the data within U, then the value at ρ of the solution always changes.

(ii) There does not exist, in fact, a larger set than Y for which property (i) holds.

We consider the above domain for the initial value problem of the same wave equation as above. The appropriate formulae for Y are given below.

The 3-dimensional case.

$$Y(t_0, x_0) = \{(t, x) \in [0, \infty) \times \mathbb{R}^3; |x - x_0| = v(t - t_0)\}.$$

The 2-dimensional case.

$$Y(t_0, x_0) = \{(t, x) \in [0, \infty) \times \mathbb{R}^2; |x - x_0| \leqslant v(t - t_0)\}.$$

Generally, we have the following theorem which is a consequence of Theorem 2.4.

THEOREM 2.7. *For the initial boundary value problem* (2.39), *the domain of influence of* $(t_0, x_0) \subset \{(t, x); |x - x_0| \leqslant v_{\max}(t - t_0)\}$ *with the provision that* $(t_0, x_0) \in [0, \infty) \times \overline{\Omega}$.

PROOF. Let $(\widetilde{t}, \widetilde{x}) \in (t_0, \infty) \times \overline{\Omega}$ be a point such that

$$(2.42) \qquad\qquad |\widetilde{x} - x_0| > v_{\max}(\widetilde{t} - t_0).$$

From Theorem 2.4, we see that the value at $(\widetilde{t}, \widetilde{x})$ of the solution u of (2.39) is determined by the value of the data within $\widetilde{\Lambda}((\widetilde{t}, \widetilde{x}), 0, \widetilde{t})$. In particular, supposition (2.42) says that $(t_0, x_0) \notin \widetilde{\Lambda}((\widetilde{t}, \widetilde{x}), 0, \widetilde{t})$. Now, if we select a neighbourhood U of (t_0, x_0) such that $U \cap \widetilde{\Lambda}((\widetilde{t}, \widetilde{x}), 0, \widetilde{t}) = \emptyset$, then no matter how the value of the data changes within U, the value at $(\widetilde{t}, \widetilde{x})$ of the solution remains unchanged. So, what this shows is that $(\widetilde{t}, \widetilde{x})$ does not belong to the domain of influence at (t_0, x_0). $\square$

(d) Finite speed of propagation.

Theorem 2.7 shows that for the phenomena governed by (2.9) the disturbance, at some point, propagates with time to its surroundings and with a "finite speed that never exceeds $v_{\max}$". Due to the above property of (2.39), that is to say, every disturbance is transmitted with finite speed, the initial boundary value problem (2.39) is often said "to have finite speed of propagation".

The finite speed of propagation, the domain of dependence and the domain of influence are concepts that are different perspectives of the same fact. The properties expressed by this nomenclature are particular properties of hyperbolic equations.

2.2 An a priori estimate of the solution

Next we focus our attention on a proof of the existence of the solution of the initial boundary value problem (2.8). First, we introduce the estimates that are necessarily satisfied when the solution does, in fact, exist. These estimates fulfill an important role in proving the existence of a solution.

In this section, in order to keep the discussion relatively uncomplicated, we assume that the coefficients of the operator P and the boundary operator B do not depend on the time variable t. We should note, however, that even if they do depend on t, the same conclusions can be derived.

(a) On the premise of results for elliptic equations.

Below, we write down some results that we would like to assume for elliptic equations. These results can be found in Lions and Magenes, [3]. In this section, we will only reference in parentheses the relevant properties and pages in [3].

First, we give a definition of the function space $H^m(\Omega)$ ($m = 0, 1, 2, \ldots$) for the solution. This space is usually called a *Sobolov space*.

DEFINITION 2.8. We say $f \in L^2(\Omega)$ belongs to $H^m(\Omega)$, if for f, thought of as a distribution, its derivatives of mth-order and below belong to $L^2(\Omega)$. If for an element f of $H^m(\Omega)$ we set

$$||f||_{m,L^2(\Omega)} = \left(\sum_{|\alpha| \leqslant m} ||\partial_x^\alpha f||_{L^2(\Omega)}^2 \right)^{1/2},$$

then $||\cdot||_{m,L^2(\Omega)}$ is the *norm* of $H^m(\Omega)$. Also, we define $H^\infty(\Omega) = \bigcap_{m=1}^{\infty} H^m(\Omega)$.

Before we go any further, we point out that $H^m(\Omega)$ is a complete metric space with distance given by the above norm.

Now, suppose that A is the elliptic operator defined by (2.5); that is, A is defined by

$$Au = \sum_{j,l=1}^{n} \frac{\partial}{\partial x_j}\left(a_{jl}\frac{\partial u}{\partial x_l}\right) - \sum_{j=1}^{n} a_j \frac{\partial u}{\partial x_j} - a_0 u,$$

and $(a_{jl})_{j,l=1,2,\ldots,n}$ satisfy (2.2). Further elaborating on this, for some $c_0 > 0$ the following is satisfied:

$$\sum_{j,l=1}^{n} a_{jl}(x)\xi_j\xi_l \geqslant c_0|\xi|^2, \qquad \forall \xi \in \mathbb{R}^n.$$

In order to distinguish between the norm of $H^{m+1/2}(\Gamma)$ ($m = 0, 1, 2, \ldots$) and the norm of $H^m(\Omega)$, denoted by $\|\cdot\|_m$, we will denote the former by $\langle \cdot \rangle_{m+1/2, L^2(\Gamma)}$.

1. An estimate of the boundary value.

$$\left\langle \left(\frac{\partial}{\partial \nu}\right)^j u \right\rangle_{l+1/2, L^2(\Gamma)} \leqslant C_m \|u\|_{j+l+1, L^2(\Omega)}.$$

([3], page 54, Théorème 9.4.)

2. An a priori estimate of the solution for elliptic equations.

For $u \in H^{m+2}(\Omega)$,

$$\|u\|_{m+2, L^2(\Omega)}$$

$$\leqslant C_m \left\{ \|Au\|_{m, L^2(\Omega)} + \left\langle \frac{\partial u}{\partial \nu_A} \right\rangle_{m+1/2, L^2(\Gamma)} + \|u\|_{m, L^2(\Omega)} \right\},$$

$$\|u\|_{m+2, L^2(\Omega)} \leqslant C_m \left\{ \|Au\|_{m, L^2(\Omega)} + \langle u \rangle_{m+3/2, L^2(\Gamma)} + \|u\|_{m, L^2(\Omega)} \right\}.$$

([3], page 166, Théorème 5.1.)

3. The existence of the solution of the boundary value problem.

With regard to the existence of the solution of the boundary value problem, namely,

$$\begin{cases} -Au + \lambda u = f & \text{in } \Omega, \\ \dfrac{\partial u}{\partial \nu_A} = g & \text{on } \Gamma, \end{cases}$$

the following is true.

THEOREM 2.9. *There exists a $\lambda_0 \in \mathbb{R}$ such that for an arbitrary $\lambda > \lambda_0$, and for $f \in L^2(\Omega)$, $g \in H^{1/2}(\Gamma)$ the solution*

$$u \in H^2(\Omega)$$

always exists and within $H^2(\Omega)$ this solution is unique. Further, for $m > 0$, if

$$f \in H^m(\Omega), \quad g \in H^{m+1/2}(\Gamma),$$

then

$$u \in H^{m+2}(\Omega).$$

(To obtain the above, we use [3], page 176, Théorème 5.4 and page 218, Corollaire 9.1.)

LEMMA 2.10. *Suppose $g \in H^\infty(\Gamma)$. For an arbitrary $\varepsilon > 0$, there exists a $u \in H^\infty(\Omega)$ such that*

$$(2.43) \qquad \frac{\partial u}{\partial \nu_A} = g, \quad ||u||_{1,L^2(\Omega)} < \varepsilon.$$

PROOF. Denote a point of Γ by s, and then write the outward unit normal vector of Γ at s as $\nu(s) = (\nu_1(s), \ldots, \nu_n(s))$. Further, we denote a neighbourhood of Γ by

$$\{x = s + r\nu(s); r \in (-\alpha, \alpha), s \in \Gamma\}.$$

Now, we take $\chi(r) \in C_0^\infty(\mathbb{R})$ such that

$$\chi(0) = 0, \quad \frac{d\chi}{dr}(0) = 1.$$

Then, for $\eta > 0$, we set

$$u_\eta(x) = \left(\sum_{j,k=1}^n a_{jk}(s)\nu_j(s)\nu_k(s) \right)^{-1} g(s)\eta\chi(r/\eta).$$

Clearly, $u_\eta \in C^\infty(\overline{\Omega}) \cap H^\infty(\Omega)$ and by construction, we have $\frac{\partial u_\eta}{\partial \nu_A} = g$. Since

$$||u_\eta||_{L^2(\Omega)} \leqslant c\eta^{3/2}, \quad \left\| \frac{\partial}{\partial x_j} u_\eta \right\|_{L^2(\Omega)} \leqslant c\eta^{1/2}$$

and by choosing $\eta > 0$ to be small, we can see that (2.43) is satisfied. □

(b) Energy estimate.

When dealing with an estimate of the solution, for the sake of conciseness we use the following notation. Suppose that $m = 1, 2, \ldots$, and for

$$u(t,x) \in \bigcap_{l=0}^{m} C^{m-l}([0,\infty); H^l(\Omega)),$$

we set

$$|||u(t)|||_{m,\Omega} = \left\{ \sum_{l=0}^{m} \left\| \left(\frac{\partial}{\partial t} \right)^{m-l} u(t,\cdot) \right\|_{l,L^2(\Omega)}^2 \right\}^{1/2}.$$

In the case $m = 1$, from the definition, (2.33), of $\tilde{e}$ and (2.2), we can see that there exists a $C_2 > 0$ such that

$$(2.44) \qquad C_2^{-1}|||u(t)|||_{1,\Omega}^2 \leqslant \int_\Omega \tilde{e}(u; t, x)dx \leqslant C_2|||u(t)|||_{1,\Omega}^2.$$

In the function space $\bigcap_{l=0}^{m} C^{m-l}([0,T]; H^l(\Omega))$ determined for each fixed $T > 0$, $\sup_{t \in [0,T]} |||u(t)|||_{m,\Omega}$ is a norm, and it is easy to confirm that this is a complete space with respect to this norm.

Now, we assume that $u \in C^2([0,\infty) \times \overline{\Omega})$ and, further, that the following is satisfied:

$$(2.45) \qquad Bu(t,x) = 0, \qquad (t,x) \in [0,\infty) \times \Gamma.$$

Also, we set $Pu(t,x) = f(t,x)$.

In Proposition 2.1 if we take $x_0 = 0$, then we can note that

$$(2.46) \qquad D(t_l) = \{x; |x| \leqslant v_{\max}(t_0 - t_l)\} \cap \Omega \qquad (l = 1, 2).$$

Next, by substituting $Bu = 0$ into (2.34), we obtain that

$$\int_{D(t_2)} \tilde{e}(u; t_2, x)dx \leqslant \int_{D(t_1)} \tilde{e}(u; t_1, x)dx$$

$$+ C \int_{t_1}^{t_2} dt \int_{D(t)} \tilde{e}(u; t, x)dx + \int_{t_1}^{t_2} \left(\int_{D(t)} |f(t,x)|^2 dx \right) dt.$$

So, if we let

$$\gamma(t) = \int_{D(t)} \tilde{e}(u; t, x)dx,$$

we have

$$\gamma(t_2) \leqslant \gamma(t_1) + C \int_{t_1}^{t_2} \gamma(t)dt + \int_{t_1}^{t_2} dt \int_{D(t)} |f(t,x)|^2 dx.$$

Since this inequality holds for all t_2 that satisfy $t_1 \leqslant t_2 < t_0$, by applying the Gronwall inequality to the above, we get that

$$\gamma(t_2) \leqslant e^{C(t_2-t_1)}\gamma(t_1) + \int_{t_1}^{t_2} e^{C(t_2-t)} \left(\int_{D(t)} |f(t,x)|^2 dx \right) dt.$$

Now, since $D(t_1) \subset \Omega$, we can see from the above that

(2.47)
$$\int_{D(t_2)} \widetilde{e}(u;t_2,x)dx \leqslant e^{C(t_2-t_1)} \int_{\Omega} \widetilde{e}(u;t_1,x)dx$$
$$+ \int_{t_1}^{t_2} e^{C(t_2-t)} \left(\int_{\Omega} |f(t,x)|^2 dx \right) dt.$$

But if t_2 is fixed and we allow $t_0 \to \infty$, then by (2.46) we have that $D(t_2) \to \Omega$; with this in mind, the implication is that

$$\int_{D(t_2)} \widetilde{e}(u;t_2,x)dx \longrightarrow \int_{\Omega} \widetilde{e}(u;t_2,x)dx.$$

Since the right-hand side of (2.47) is independent of t_0, the following result holds as a consequence of (2.44).

PROPOSITION 2.11. *Suppose that $u \in C^2([0,\infty) \times \overline{\Omega})$ satisfies the boundary condition. Then, for arbitrary $0 \leqslant t_1 < t_2$,*

(2.48)
$$|||u(t_2)|||_{1,\Omega}^2$$
$$\leqslant C_1 \left\{ e^{C(t_2-t_1)} |||u(t_1)|||_{1,\Omega}^2 + \int_{t_1}^{t_2} e^{C(t_2-t)} ||f(t,\cdot)||_{L^2(\Omega)}^2 dt \right\},$$

where C and C_1 are constants independent of u and t_1, t_2.

(c) **An estimate of partial derivatives of higher order.**
    We derive an estimate that holds for the smooth solution u of the initial boundary value problem (2.8).

PROPOSITION 2.12. *Suppose that u is a solution (2.8) and that u satisfies*

$$\left(\frac{\partial}{\partial t}\right)^l u \in C^2([0,\infty) \times \overline{\Omega}) \qquad (l = 0, 1, 2, \ldots, m+3).$$

If f, $\dfrac{\partial f}{\partial t} \in \displaystyle\bigcap_{l=0}^{m} C^{m-l}([0,\infty); H^l(\Omega))$, then

(2.49)
$$u \in \bigcap_{l=0}^{m+2} C^{m+2-l}([0,\infty); H^l(\Omega)),$$

and, further, for arbitrary $0 \leqslant t_1 < t$ the following inequality holds:
(2.50)
$$|||u(t)|||_{m+2,\Omega}^2 \leqslant C_m \left\{ e^{C(t-t_1)} |||u(t_1)|||_{m+2,\Omega}^2 + |||f(t_1)|||_{m,\Omega}^2 \right.$$
$$\left. + (t - t_1 + 1) \int_{t_1}^{t} e^{C(t-s)} \left|\left|\left|\frac{\partial f}{\partial t}(s)\right|\right|\right|_{m,\Omega}^2 ds \right\}.$$

For every $t \geqslant 0$, the following estimate also holds true:
(2.51)
$$C_m^{-1} |||u(t)|||_{m+2,\Omega}^2 \leqslant \|u(t,\cdot)\|_{m+2,L^2(\Omega)}^2 + \left\|\frac{\partial u}{\partial t}(t,\cdot)\right\|_{m+1,L^2(\Omega)}^2$$
$$+ |||f(t)|||_{m,\Omega}^2$$
$$\leqslant C_m |||u(t)|||_{m+2,\Omega}^2.$$

We first prove (2.51). In order to show the right-hand part of the inequality, it is sufficient to show that $|||f(t)|||_{m,\Omega} \leqslant C|||u(t)|||_{m+2,\Omega}$. However, this follows directly from

(2.52)
$$f = \frac{\partial^2 u}{\partial t^2} + H\frac{\partial u}{\partial t} - Au.$$

Next, we show the left-hand part of the inequality. Using

(2.53)
$$\frac{\partial^2 u}{\partial t^2} = -H\frac{\partial u}{\partial t} + Au + f,$$

we obtain that
(2.54)

$$\left\|\frac{\partial^2 u}{\partial t^2}(t,\cdot)\right\|^2_{m,L^2(\Omega)} \leqslant C\left\{\left\|\frac{\partial u}{\partial t}(t,\cdot)\right\|^2_{m+1,L^2(\Omega)} + \|u(t,\cdot)\|^2_{m+2,L^2(\Omega)}\right.$$
$$\left. + \|f(t,\cdot)\|^2_{m,L^2(\Omega)}\right\}.$$

For the case $m \geqslant 1$, by partially differentiating with respect to t both sides of (2.53), we see that

$$\left\|\frac{\partial^3 u}{\partial t^3}(t,\cdot)\right\|^2_{m-1,L^2(\Omega)} \leqslant C\left\{\left\|\frac{\partial^2 u}{\partial t^2}(t,\cdot)\right\|^2_{m,L^2(\Omega)}\right.$$
$$\left. + \left\|\frac{\partial u}{\partial t}(t,\cdot)\right\|^2_{m+1,L^2(\Omega)} + \left\|\frac{\partial f}{\partial t}(t,\cdot)\right\|^2_{m-1,L^2(\Omega)}\right\}.$$

Now, if we substitute (2.54) into the right-hand side of the above, then by making use of the middle term of (2.51) we can estimate

$$\left\|\frac{\partial^3 u}{\partial t^3}(t,\cdot)\right\|^2_{m-1,L^2(\Omega)}.$$

By repeating the above, we eventually derive the left-hand part of the inequality of (2.51). To this end, we begin with the following lemma.

LEMMA 2.13. *For* $m = 1, 2, \ldots$ *there exists a constant* $C_m > 0$ *which is independent of* u, *and such that the following holds:*

$$C_m^{-1}\||u(t)\||^2_{m+1,\Omega}$$
(2.55)
$$\leqslant \left\{\left\|\frac{\partial u}{\partial t}(t)\right\|^2_{m,\Omega} + \||u(t)\||^2_{m,\Omega} + \||f(t)\||^2_{m-1,\Omega}\right\}$$
$$\leqslant C_m\||u(t)\||^2_{m+1,\Omega}.$$

PROOF. In order to prove the left-hand part of the inequality, it is sufficient to show that $\|u(t,\cdot)\|_{m+1,L^2(\Omega)}$ can be estimated by the right-hand side. From (2.8), we can write

$$\begin{cases} Au(t,x) = \dfrac{\partial^2 u}{\partial t^2}(t,x) + H\dfrac{\partial u}{\partial t}(t,x), \\ \dfrac{\partial u}{\partial \nu_A}(t,x) = \sigma_0(x)\dfrac{\partial u}{\partial t}(t,x). \end{cases}$$

Now, if we use the *a priori* estimate of the elliptic operator discussed in the previous section, then we obtain that

$$\|u(t,\cdot)\|^2_{m+1,L^2(\Omega)}$$

$$\leqslant C_m \left\{ \left\|\frac{\partial^2 u}{\partial t^2}(t,\cdot)\right\|^2_{m-1,L^2(\Omega)} + \left\|H\frac{\partial u}{\partial t}(t,\cdot)\right\|^2_{m-1,L^2(\Omega)} \right.$$

$$\left. +\|f(t,\cdot)\|^2_{m-1,L^2(\Omega)} + \left\langle \sigma_0 \frac{\partial u}{\partial t}(t,\cdot)\right\rangle^2_{m-1+1/2,L^2(\Gamma)} \right\}.$$

Using the following estimate of the boundary value,

$$\left\langle \sigma_0 \frac{\partial u}{\partial t}(t,\cdot)\right\rangle^2_{m-1+1/2,L^2(\Gamma)} \leqslant C_m \left\|\frac{\partial u}{\partial t}\right\|^2_{m,L^2(\Omega)},$$

it is easy to see that $\|u(t)\|_{m+1,L^2(\Omega)}$ can be estimated by the middle term in (2.55).

For the right-hand part of the inequality, it is sufficient for us to show that $\|\|f(t)\|\|_{m-1,\Omega}$ can be estimated by $C\|\|u(t)\|\|_{m+1,\Omega}$. But, this follows immediately from (2.52).

$\square$

The above preparations now allow us to begin a proof of Proposition 2.12.

To start, suppose $\dfrac{\partial u}{\partial t} \in C^2([0,\infty) \times \overline{\Omega})$. An easy consequence of partially differentiating (2.45) with respect to t is

$$\left(B\frac{\partial u}{\partial t}\right)(t,x) = 0, \qquad (t,x) \in [0,\infty) \times \Gamma.$$

Also, by partially differentiating $Pu = f$ again with respect to t, we have that

$$P\frac{\partial u}{\partial t} = \frac{\partial f}{\partial t}.$$

Therefore, from the above, we can apply Proposition 2.11 to $\dfrac{\partial u}{\partial t}$, and hence we see that

(2.56)
$$\left\|\left\|\frac{\partial u}{\partial t}(t_2)\right\|\right\|^2_{1,\Omega} \leqslant C_1 \left\{ e^{C(t_2-t_1)}\left\|\left\|\frac{\partial u}{\partial t}(t_1)\right\|\right\|^2_{1,\Omega} \right.$$

$$\left. + \int_{t_1}^{t_2} e^{C(t_2-t)}\left\|\frac{\partial f}{\partial t}(t,\cdot)\right\|^2_{L^2(\Omega)} dt \right\}.$$

If we now apply Lemma 2.13 with $m = 1$, then we obtain
(2.57)
$$|||u(t)|||_{2,\Omega}^2 \leqslant C \left\{ \left|\left|\left| \frac{\partial u}{\partial t}(t) \right|\right|\right|_{1,\Omega}^2 + |||u(t)|||_{1,\Omega}^2 + ||f(t,\cdot)||_{L^2(\Omega)}^2 \right\}.$$

Taking $t = t_2$, and substituting (2.48), (2.56) into the right-hand side of (2.57), we find that
(2.58)
$$|||u(t_2)|||_{2,\Omega}^2$$
$$\leqslant C_3 \left\{ e^{C(t_2-t_1)} \left(\left|\left|\left| \frac{\partial u}{\partial t}(t_1) \right|\right|\right|_{1,\Omega}^2 + |||u(t_1)|||_{1,\Omega}^2 \right) \right.$$
$$\left. + \int_{t_1}^{t_2} e^{C(t_2-t)} \left(||f(t,\cdot)||_{L^2(\Omega)}^2 + \left|\left| \frac{\partial f}{\partial t}(t,\cdot) \right|\right|_{L^2(\Omega)}^2 \right) dt \right\}.$$

So, for $t \geqslant t_1$ since
$$||f(t,\cdot)||_{L^2(\Omega)}^2$$
$$\leqslant 2 \left\{ ||f(t_1,\cdot)||_{L^2(\Omega)}^2 + (t - t_1) \int_{t_1}^t \left|\left| \frac{\partial f}{\partial t}(s,\cdot) \right|\right|_{L^2(\Omega)}^2 ds \right\},$$

and assuming that $C_1 \geqslant 1$, we obtain that
$$\int_{t_1}^{t_2} e^{C(t_2-t)} \left(||f(t,\cdot)||_{L^2(\Omega)}^2 + \left|\left| \frac{\partial f}{\partial t}(t,\cdot) \right|\right|_{L^2(\Omega)}^2 \right) dt$$
(2.59)
$$\leqslant 2e^{C(t_2-t_1)} ||f(t_1,\cdot)||_{L^2(\Omega)}^2$$
$$+ \int_{t_1}^{t_2} \{1 + 2(t_2 - t)\} e^{C(t_2-t)} \left|\left| \frac{\partial f}{\partial t}(t,\cdot) \right|\right|_{L^2(\Omega)}^2 dt.$$

Now, for arbitrary $0 \leqslant t_1 < t_2$, substituting (2.59) into the right-hand side of (2.58) and then with $t = t_1$ by applying the right-hand inequality of (2.51) to the right-hand side, we see that
(2.60)
$$|||u(t_2)|||_{2,\Omega}^2 \leqslant C_2 \left\{ e^{C(t_2-t_1)} |||u(t_1)|||_{2,\Omega}^2 \right.$$
$$\left. + \int_{t_1}^{t_2} (1 + t_2 - t) e^{C(t_2-t)} \left|\left| \frac{\partial f}{\partial t}(t,\cdot) \right|\right|_{L^2(\Omega)}^2 dt \right\}.$$

Hence, (2.50) is proved for the case of $m = 0$.

Now, assume that $\dfrac{\partial^2 u}{\partial t^2} \in C^2([0,\infty) \times \overline{\Omega})$. Since we can use the results of the previous steps on $\dfrac{\partial u}{\partial t}$, when $t \geqslant t_1$ we get that

(2.61)
$$\left|\left|\left|\frac{\partial u}{\partial t}(t)\right|\right|\right|^2_{2,\Omega} \leqslant C_5 \left\{ e^{C(t-t_1)} \left|\left|\left|\frac{\partial u}{\partial t}(t_1)\right|\right|\right|^2_{2,\Omega} \right.$$
$$\left. + \int_{t_1}^t (1+t-s)e^{C(t-s)} \left|\left|\frac{\partial^2 f}{\partial t^2}(s,\cdot)\right|\right|^2_{L^2(\Omega)} ds \right\}.$$

By using (2.55) with $m = 2$ and applying (2.61), we obtain that

(2.62)
$$|||u(t)|||^2_{3,\Omega} \leqslant C_3 \left\{ e^{C(t-t_1)} \left|\left|\left|\frac{\partial u}{\partial t}(t_1)\right|\right|\right|^2_{2,\Omega} \right.$$
$$+ \int_{t_1}^t (1+t-t_1)e^{C(t-s)} \left|\left|\frac{\partial^2 f}{\partial t^2}(s,\cdot)\right|\right|^2_{L^2(\Omega)} ds$$
$$\left. + |||u(t)|||^2_{2,\Omega} + |||f(t)|||^2_{1,\Omega} \right\}.$$

Note that

(2.63)
$$||f(t)||^2_{1,L(\Omega)}$$
$$\leqslant 2 \left\{ ||f(t_1,\cdot)||^2_{1,L(\Omega)} + (t-t_1) \int_{t_1}^t \left|\left|\frac{\partial f}{\partial t}(s,\cdot)\right|\right|^2_{1,L^2(\Omega)} ds \right\}.$$

From the above, by substituting (2.63) into the right-hand side of (2.62) and by estimating $|||u(t)|||_{2,\Omega}$ by means of (2.60), we have that

(2.64)
$$|||u(t)|||^2_{3,\Omega} \leqslant C_3 \left\{ e^{C(t-t_1)} \left|\left|\left|\frac{\partial u}{\partial t}(t_1)\right|\right|\right|^2_{2,\Omega} + |||f(t_1)|||^2_{1,\Omega} \right.$$
$$\left. + (1+t-t_1) \int_{t_1}^t e^{C(t-s)} \left|\left|\left|\frac{\partial f}{\partial t}(s)\right|\right|\right|^2_{1,\Omega} ds \right\}.$$

Now, in (2.64) we use the right-hand part of the inequality in (2.55), and if necessary we replace the constant C_3 by something larger; then we can easily see that the following holds:

$$\||u(t)\||^2_{3,\Omega} \leqslant C_3 \left\{ e^{C(t-t_1)} \||u(t_1)\||^2_{3,\Omega} \right.$$

$$\left. + (t - t_1 + 1) \int_{t_1}^t e^{C(t-s)} \left\|\left|\left|\frac{\partial f}{\partial t}(s)\right|\right|\right\|^2_{1,\Omega} ds \right\}.$$

So, by repeating the above procedure, we obtain Proposition 2.12.

(d) Elimination of the extra assumption, approximation due to a mollifier.

In Proposition 2.12, in order to obtain an estimate of the solution, we had to make an extra assumption concerning the smoothness of the solution. That is to say, in order to show (2.50), it was required for u to be more intrinsically strongly smooth than the smoothness of the solution estimated by (2.50). In this section, we would like to consider an approach to (2.50) without the extra assumption about smoothness. For this to work, we apply a method that approximates u by a sequence of sufficiently smooth functions which rely on a so-called mollifier.

Suppose that ρ is a non-negative function defined on $\mathbb{R}$ that satisfies the following:

$$(2.65) \qquad \rho(l) \in C^\infty(\mathbb{R}), \quad \operatorname{supp} \rho \subset [-1, 1] \text{ and } \int_{-\infty}^\infty \rho(l)dl = 1.$$

For $\varepsilon > 0$, we set

$$\rho_\varepsilon(l) = \varepsilon^{-1} \rho\left(\frac{l}{\varepsilon} + 1\right).$$

Hence, from the above, the following holds:

$$(2.66) \qquad \operatorname{supp} \rho_\varepsilon \subset [-2\varepsilon, 0] \text{ and } \int_{-\infty}^\infty \rho_\varepsilon(l)dl = 1.$$

For u such that

$$(2.67) \qquad u \in \bigcap_{j=0}^{m+2} C^{m+2-j}([0,\infty); H^j(\Omega)),$$

we set

$$(2.68) \qquad u_\varepsilon(t,x) = (\rho_\varepsilon * u)(t,x) = \int_{-\infty}^\infty \rho_\varepsilon(t-s)u(s,x)ds.$$

We call $\rho_\varepsilon *$ the *mollifier*.

Let us now suppose that $t \geqslant 0$; then the above integral in s becomes an integral over the interval $[t, t + 2\varepsilon]$. For $t \geqslant 0$, we can write

(2.69)
$$\left(\frac{\partial}{\partial t}\right)^p u_\varepsilon(t, x) = \int_{-\infty}^{\infty} \frac{\partial^p \rho_\varepsilon(t - s)}{\partial t^p} u(s, x) ds \qquad (p = 1, 2, \ldots).$$

Using the relation
$$\frac{\partial^p \rho_\varepsilon(t - s)}{\partial t^p} = (-1)^p \frac{\partial^p \rho_\varepsilon(t - s)}{\partial s^p},$$

and integrating by parts for $p \leqslant m + 2$, we obtain that

(2.70)
$$\left(\frac{\partial}{\partial t}\right)^p u_\varepsilon(t, x) = \int_{-\infty}^{\infty} \rho_\varepsilon(t - s) \frac{\partial^p u}{\partial s^p}(s, x) ds = \left(\rho_\varepsilon * \frac{\partial^p u}{\partial t^p}\right)(t, x).$$

PROPOSITION 2.14. *We assume* (2.67). *For $p \leqslant m + 2$ and with $t \geqslant 0$, if we let $\varepsilon \to 0$, then the following is true:*

(2.71) $$\left(\frac{\partial}{\partial t}\right)^p u_\varepsilon(t, \cdot) \longrightarrow \left(\frac{\partial}{\partial t}\right)^p u(t, \cdot) \qquad in \ H^{m+2-p}(\Omega).$$

PROOF. From the assumption, $\dfrac{\partial^p u}{\partial t^p}(t, \cdot)$ is continuous with respect to the topology of $H^{m+2-p}(\Omega)$. So, using $\displaystyle\int \rho_\varepsilon(t - s) ds = 1$, we have that

$$\left(\frac{\partial}{\partial t}\right)^p u_\varepsilon(t, \cdot) - \left(\frac{\partial}{\partial t}\right)^p u(t, \cdot)$$
$$= \int_{-\infty}^{\infty} \rho_\varepsilon(t - s) \left\{ \frac{\partial^p u}{\partial t^p}(s, \cdot) - \frac{\partial^p u}{\partial t^p}(t, \cdot) \right\} ds.$$

Since the integral of the right-hand side is on the interval $[t, t+2\varepsilon]$, we get

$$\left\| \left(\frac{\partial}{\partial t}\right)^p u_\varepsilon(t, \cdot) - \left(\frac{\partial}{\partial t}\right)^p u(t, \cdot) \right\|_{m+2-p, L^2(\Omega)}$$
$$\leqslant \sup_{s \in [t, t+2\varepsilon]} \left\| \frac{\partial^p u}{\partial t^p}(s, \cdot) - \frac{\partial^p u}{\partial t^p}(t, \cdot) \right\|_{m+2-p, L^2(\Omega)} \int_{-\infty}^{\infty} \rho_\varepsilon(t - s) ds,$$

and then from the continuity of $\dfrac{\partial^p u}{\partial t^p}$, the right-hand side converges to zero as $\varepsilon \to 0$.

$\square$

Since we have assumed that the coefficients of P are independent of t, making use of the mollifier $\rho_\varepsilon *$ in both sides of $Pu = f$, and then using (2.70), for $t \geqslant 0$, we see that

$$(2.72)\qquad f_\varepsilon(t,x) = (\rho_\varepsilon * Pu)(t,x) = P(\rho_\varepsilon * u)(t,x) = (Pu_\varepsilon)(t,x).$$

In a similar fashion we can confirm that

$$(2.73)\qquad (Bu_\varepsilon)(t,x) = 0 \qquad ((t,x) \in [0,\infty) \times \Gamma).$$

Now, from (2.69), we have that

$$u_\varepsilon(t,\cdot) \in C^\infty([0,\infty); H^{m+2}(\Omega)).$$

Therefore, by making use of Proposition 2.12, we can show that the following holds:
(2.74)

$$|||u_\varepsilon(t)|||^2_{m+2,\Omega} \leqslant C_m \left\{ e^{C(t-t_1)}(|||u_\varepsilon(t_1)|||^2_{m+2,\Omega} + |||f_\varepsilon(t_1)|||^2_{m,\Omega}) \right.$$

$$\left. + (t - t_1 + 1)\int_{t_1}^t e^{C(t-s)} \left|\left|\left| \frac{\partial f_\varepsilon}{\partial t}(s) \right|\right|\right|^2_{m,\Omega} ds \right\}.$$

From Proposition 2.14 with $t \geqslant 0$ and letting $\varepsilon \to 0$, we can show that

$$|||u_\varepsilon(t)|||_{m+2,\Omega} \longrightarrow |||u(t)|||_{m+2,\Omega},$$

$$|||f_\varepsilon(t)|||_{m,\Omega} \longrightarrow |||f(t)|||_{m,\Omega},$$

$$\left|\left|\left| \frac{\partial f_\varepsilon}{\partial t}(t) \right|\right|\right|_{m,\Omega} \longrightarrow \left|\left|\left| \frac{\partial f}{\partial t}(t) \right|\right|\right|_{m,\Omega}.$$

Hence, from the above and by allowing $\varepsilon \to 0$ in (2.74), we can deduce the following theorem.

THEOREM 2.15. *Assume u is a solution of (2.8), and u satisfies the following:*

$$u \in \bigcap_{l=0}^{m+2} C^{m+2-l}([0,\infty); H^l(\Omega)).$$

If

$$f, \ \frac{\partial f}{\partial t} \in \bigcap_{l=0}^{m} C^{m+1-l}([0,\infty); H^l(\Omega)),$$

then for $0 \leqslant t_1 < t$ the following holds:

$$|||u(t)|||_{m+2,\Omega}^2 \leqslant C_m \left\{ e^{C(t-t_1)}(|||u(t_1)|||_{m+2,\Omega}^2 + |||f(t_1)|||_{m,\Omega}^2) \right.$$

$$\left. +(t - t_1 + 1) \int_{t_1}^{t} e^{C(t-s)} \left|\left|\left| \frac{\partial f}{\partial t}(s) \right|\right|\right|_m^2 \, ds \right\}.$$

Also, for $|||u(t_1)|||_{m+2,\Omega}$ in the right-hand side, the following estimate is true:

$$|||u(t_1)|||_{m+2,\Omega}^2 \leqslant C_m \left\{ ||u(t_1,\cdot)||_{m+2,L^2(\Omega)}^2 + \left|\left| \frac{\partial u}{\partial t}(t_1,\cdot) \right|\right|_{m+1,L^2(\Omega)}^2 \right.$$

$$\left. + |||f(t_1)|||_{m,\Omega}^2 \right\}.$$

(e) Energy conservation.

We consider the hyperbolic operator P with the form

(2.75)
$$Pu = \frac{\partial^2 u}{\partial t^2} - \sum_{j,k=1}^{n} \frac{\partial}{\partial x_j}\left(a_{jk}(x) \frac{\partial u}{\partial x_k} \right),$$

where the coefficients are independent of the time variable t. Further, suppose that $(a_{jk})_{j,k=1,2,\dots,n}$ are elliptic; namely, satisfy (2.2).

Using P of the above form and from the calculations in §2.1, we have

$$2\operatorname{Re} Pu \overline{\frac{\partial u}{\partial t}} = \frac{\partial}{\partial t}\left(\left|\frac{\partial u}{\partial t}\right|^2 + \sum_{j,k=1}^{n} a_{jk} \frac{\partial u}{\partial x_j} \overline{\frac{\partial u}{\partial x_k}} \right)$$

$$- 2\operatorname{Re} \sum_{j,k=1}^{n} \frac{\partial}{\partial x_j}\left(a_{jk} \frac{\partial u}{\partial x_k} \overline{\frac{\partial u}{\partial t}} \right).$$

Therefore, (2.21) becomes

(2.76)

$$\int_V 2\mathrm{Re}\, Pu\overline{\frac{\partial u}{\partial t}}\, dt dx = \int_{D(t_2)} e(u; t_2, x) dx$$
$$- \int_{D(t_1)} e(u; t_1, x) dx - \int_{S_b} \sum_{j=1}^{n} X_j(u; t, x)\nu_j dS$$
$$+ \int_S \left(e(u; t, x)\eta + \sum_{j=1}^{n} X_j(u; t, x)\mu_j \right) dS.$$

Recall that we showed in §2.1 that the integral over S is non-negative.

Now take the boundary operator B as $B = \dfrac{\partial}{\partial \nu_A}$, which in turn satisfies condition (2.11).

If $u \in H^2(\mathbb{R} \times \Omega)$ satisfies

(2.77)
$$\begin{cases} Pu(t, x) = 0 & \text{in } \mathbb{R} \times \Omega, \\ Bu = 0 & \text{on } \mathbb{R} \times \Gamma, \end{cases}$$

then

$$\int_{D(t_2)} e(u; t_2, x) dx \leqslant \int_{D(t_1)} e(u; t_1, x) dx.$$

Since $D(t_1) \subset \Omega$,

$$\int_{D(t_2)} e(u; t_2, x) dx \leqslant \int_\Omega e(u; t_1, x) dx.$$

If we allow $t_0 \to \infty$, where t_0 is an indeterminant that appears in the definition of V, then $D(t_2)$ converges to Ω. Hence,

(2.78)
$$\int_\Omega e(u; t_2, x) dx \leqslant \int_\Omega e(u; t_1, x) dx.$$

Next, we set $\Lambda^+(t_0, x_0) = \{(t, x); |x - x_0| \leqslant v_{\max}(t - t_0)\}$. For $t_0 < t_1 < t_2$, write

$$V^+(t_1, t_2) = \Lambda^+(t_0, x_0) \cap ([t_1, t_2] \times \Omega).$$

Since condition (2.11) is satisfied even when, as in this case, we take time in the reverse direction, so, by a similar calculation, we obtain that

$$\int_{D^+(t_1)} e(u; t_1, x) dx \leqslant \int_{D^+(t_2)} e(u; t_2, x) dx,$$

where

$$D^+(\tau) = \{(\tau, x); |x - x_0| \leqslant v_{\max}(\tau - t_0), x \in \Omega\}.$$

Then, if we allow $t_0 \to -\infty$, we see that

(2.79)
$$\int_\Omega e(u; t_1, x)dx \leqslant \int_\Omega e(u; t_2, x)dx.$$

Hence, from (2.78) and (2.79), for arbitrary $t_1 < t_2$, we have that

(2.80)
$$\int_\Omega e(u; t_1, x)dx = \int_\Omega e(u; t_2, x)dx.$$

Let

$$E(u; t) = \frac{1}{2} \int_\Omega e(u; t, x)dx;$$

then $E(u, t)$ is said to be the *total energy* of the solution u at time t, and (2.80) gives the next theorem.

THEOREM 2.16. *Suppose that the hyperbolic operator P has the form given in (2.75) and $B = \dfrac{\partial}{\partial \nu_A}$. If $u \in H^2(\mathbb{R} \times \Omega)$ satisfies*

$$\begin{cases} Pu = 0 & in \ \mathbb{R} \times \Omega, \\ Bu = 0 & on \ \mathbb{R} \times \Gamma, \end{cases}$$

then the total energy of the solution is conserved.

2.3 Existence of the solution

(a) The Hille-Yosida theorem.

In order to prove the existence of the solution of the initial boundary value problem (2.8) we need to make use of a theorem usually referred to as the Hille-Yosida theorem, which deals with semi-groups of operators, for instance, see Yosida [2], Chapter IX, Section 7. For our purposes, we need only the part of the Hille-Yosida theorem that discusses the formation of a semi-group for a generating operator A using an assumption on the resolvent of the operator A, and so we shall only give a proof of that part. Since we require only a partial version of the Hille-Yosida theorem, the title of this section is not strictly correct, but we are sure that this will not cause any concern. In spite of the above, we will refer to Theorem 2.18 as the Hille-Yosida theorem. Also, we would like to forewarn the reader, but there should really not be any confusion, that the A in this section is different from the A given by (2.5).

For the purposes of this section, we will assume that H is a Hilbert space, and we denote an element of H by x. In the standard way, we shall denote the inner product of H by $(\cdot, \cdot)$ and the norm by $||\cdot||$. Further, we assume that A is a linear operator that has a dense domain $D(A)$, and there exists an $M > 0$ such that the following is satisfied:

$$(2.81) \qquad 2\mathrm{Re}\,(Ax, x) \leqslant M||x||^2, \qquad \forall x \in D(A).$$

Further, we assume there is a $\lambda_0 \in \mathbb{R}$ such that for arbitrary $\lambda \geqslant \lambda_0$,

$$(2.82) \qquad\qquad (\lambda - A)^{-1} \text{ exists.}$$

LEMMA 2.17. *For* $\lambda > \max(\lambda_0, M)$ $(= \lambda_1)$ *the following estimate holds:*

$$(2.83) \qquad\qquad ||(\lambda - A)^{-1}|| \leqslant (\lambda - M)^{-1}.$$

PROOF. For $x \in D(A)$,

$$((\lambda - A)x, (\lambda - A)x) = \lambda^2 ||x||^2 - 2\lambda\mathrm{Re}\,(Ax, x) + ||Ax||^2$$
$$\geqslant (\lambda^2 - \lambda M)||x||^2.$$

If $\lambda \geqslant M$, from the fact that $\lambda^2 - \lambda M \geqslant (\lambda - M)^2$, we obtain

$$(2.84) \qquad\qquad ||(\lambda - A)x||^2 \geqslant (\lambda - M)^2 ||x||^2.$$

This now shows (2.83).

$\square$

For $\lambda \geqslant \lambda_1 + 1$ let

$$J_\lambda = \lambda(\lambda - A)^{-1} = (I - \lambda^{-1}A)^{-1}.$$

From (2.83), we get that

$$(2.85) \qquad ||J_\lambda|| \leqslant \lambda(\lambda - M)^{-1} = (1 - \lambda^{-1}M)^{-1}.$$

If $x \in D(A)$, then since $x - J_\lambda x = -(\lambda - A)^{-1}Ax$, when $\lambda \to \infty$ we see that $J_\lambda x \to x$. Since $\{J_\lambda\}_{\lambda \geqslant \lambda_1 + 1}$ is bounded as a set of operators of H, then by using this with the fact that $D(A)$ is dense, we obtain that

$$(2.86) \qquad J_\lambda x \longrightarrow x \quad (\lambda \longrightarrow \infty), \qquad \forall x \in H.$$

Now, since

$$(2.87) \qquad\qquad AJ_\lambda = \lambda J_\lambda - \lambda,$$

for each fixed λ, we should note that AJ_λ becomes a bounded operator in H. Therefore, for $t \geqslant 0$ we can define the operator $U_\lambda(t)$ in H as

$$(2.88) \qquad U_\lambda(t) = \exp(tAJ_\lambda) = I + tAJ_\lambda + \frac{1}{2!}(tAJ_\lambda)^2 + \cdots.$$

Using (2.87), we can express $U_\lambda(t)$ as

$$U_\lambda(t) = \exp(t\lambda J_\lambda - t\lambda) = \exp(t\lambda J_\lambda)\exp(-t\lambda).$$

Since $\|\exp(t\lambda J_\lambda)\| \leqslant \exp(t\lambda\|J_\lambda\|)$ and further because

$$\|U_\lambda(t)\| \leqslant \exp(t\lambda(1 - \lambda^{-1}M)^{-1})\exp(-t\lambda) \leqslant \exp(tM(1 - \lambda^{-1}M)^{-1}),$$

then there exists a constant $C > 0$ such that

$$(2.89) \qquad \|U_\lambda(t)\| \leqslant \exp(Ct), \qquad \forall t \geqslant 0, \ \forall \lambda \geqslant \lambda_1 + 1.$$

Also, as a property of exponential functions, for $t, s \geqslant 0$, the following hold:

$$(2.90) \ \ U_\lambda(t+s) = U_\lambda(t)U_\lambda(s), \quad \frac{dU_\lambda(t)}{dt} = (AJ_\lambda)U_\lambda(t) = U_\lambda(t)AJ_\lambda.$$

Next, we show the convergence of $U_\lambda(t)$ as $\lambda \to \infty$. For this purpose, we assume that $\mu > \lambda_1 + 1$. From the fact that $J_\mu J_\lambda = J_\lambda J_\mu$ we see that $U_\mu(s)J_\lambda = J_\lambda U_\mu(s)$. On the other hand, since

$$\frac{d}{ds}\{U_\lambda(t-s)U_\mu(s)\} = U_\lambda(t-s)(-AJ_\lambda + AJ_\mu)U_\mu(s),$$

under the condition $x \in D(A)$ we can integrate both sides over the interval $[0, t]$. Since the operators in the right-hand side mutually commute, we obtain the following expression:

$$U_\mu(t)x - U_\lambda(t)x = \int_0^t U_\lambda(t-s)U_\mu(s)(J_\mu - J_\lambda)Ax\,ds.$$

Using (2.89), we see that the integrand of the right-hand side of the above is estimated by $\exp(Ct)\|(J_\mu - J_\lambda)Ax\|$ for all $s \in [0, t]$. Therefore, from (2.86) as $\lambda, \mu \to \infty$ the integral of the right-hand side converges to zero. This implies that for $x \in D(A)$ as $\lambda \to \infty$ the term $U_\lambda(t)x$ converges to some element in H. Now, for $x \in D(A)$ we define $U(t)$ by

$$U(t)x = \lim_{\lambda\to\infty} U_\lambda(t)x.$$

From (2.89), $\{U_\lambda(t)\}_{\lambda \geqslant \lambda_1 + 1}$ is uniformly bounded and since $D(A)$ is dense, for arbitrary $x \in H$ we obtain the existence of $U(t)x = \lim_{\lambda \to \infty} U_\lambda(t)x$. The next set of properties follows directly from the properties of $U_\lambda(t)$:

$$(2.91) \qquad \qquad \|U(t)x\| \leqslant \exp(Ct)\|x\|,$$

$$(2.92) \qquad \qquad U(t+s) = U(t)U(s),$$

$$(2.93) \qquad \qquad U(0) = I.$$

Also, $U_\lambda(t)x$ is continuous at $t \geqslant 0$ and if we fix $x \in D(A)$, then, for any bounded interval of t, $U_\lambda(t)x$ converges uniformly to $U(t)x$. Therefore, $U(t)x$ is continuous at $t \geqslant 0$. Now, combining the fact that $D(A)$ is dense with (2.91), we obtain that

(2.94) \qquad for each $x \in H$, $U(t)x$ is continuous at $t \geqslant 0$.

Now, if we integrate with respect to t both sides of the second formula in (2.90) over the interval $[0, t]$, we get

$$U_\lambda(t)x - x = \int_0^t U_\lambda(s) A J_\lambda x \, ds.$$

But, if we assume that $x \in D(A)$, then by (2.86) and the definition of $U(t)$, as $\lambda \to \infty$ we can deduce that

$$U(t)x = x + \int_0^t U(s) Ax \, ds.$$

Since $U(s)Ax$ is a continuous function, it is easy to see that the right-hand side can be differentiated with respect to t, and, further, $U(t)x$ is differentiable at $t \geqslant 0$. So,

$$(2.95) \qquad \frac{dU(t)x}{dt} = U(t)Ax = AU(t)x.$$

Collecting the above together leads to the following theorem.

THEOREM 2.18 (Hille-Yosida theorem). *Suppose* (2.81), (2.82) *are satisfied by an operator A in the Hilbert space H. Then, there exists a family $\{U(t)\}$ of bounded operators of H on the parameter $t \geqslant 0$ with the following properties:*
(i) *For arbitrary t, $s \in [0, \infty)$,*

$$U(t)U(s) = U(t+s), \quad U(0) = I.$$

(ii) *For all $x \in H$, $U(t)x$ is continuous with respect to $t \in [0, \infty)$ in the strong topology of H.*

(iii) *If $x \in D(A)$, $U(t)x$ is continuously differentiable with respect to $t \in (0, \infty)$ in the strong topology and satisfies (2.95).*

COROLLARY 2.19. *Suppose that $f(t) \in D(A)$ and, further, $f(t)$, $Af(t) \in C([0, \infty); H)$. If $x \in D(A)$, let us set*

$$x(t) = U(t)x + \int_0^t U(t-s)f(s)ds.$$

Then $x(t)$ is continuously differentiable in the strong topology at $(0, \infty)$ and satisfies

$$\begin{cases} \dfrac{dx(t)}{dt} = Ax(t) + f(t) & (t \in (0, \infty)), \\ x(0) = x. \end{cases}$$

(b) The operator $\mathcal{A}$.

We once more write down the problem for which we want to show the existence of the solution.

(2.96)
$$\begin{cases} Pu = f & \text{in } (0, \infty) \times \Omega, \\ Bu = 0 & \text{in } (0, \infty) \times \Gamma, \\ u(0, x) = u_0(x), \ \dfrac{\partial u}{\partial t}(0, x) = u_1(x). \end{cases}$$

In order to show the existence of the solution we will make use of the existence theorem, proven in (a), for the solution of an evolution equation. For us to be able to use the Hille-Yosida theorem, we need to suitably select the Hilbert space and the operator in problem (2.96). So, with this in mind we denote the Hilbert space we need here by $\mathcal{H}$ and an operator in it by $\mathcal{A}$. Therefore, let the symbols A and H which are partial differential operators defined respectively by (2.5) and (2.6) be understood in this manner.

Now, we take the Hilbert space $\mathcal{H}$ to be

(2.97) $\mathcal{H} = H^1(\Omega) \times L^2(\Omega).$

Then, for elements $X = \{v_0, v_1\}$, $Y = \{w_0, w_1\}$ of $\mathcal{H}$, where $v_0, w_0 \in H^1(\Omega)$ and $v_1, w_1 \in L^2(\Omega)$, we define the inner product $(\cdot, \cdot)_{\mathcal{H}}$ by
(2.98)
$$(X, Y)_{\mathcal{H}} = \sum_{j,l=1}^n \int_\Omega a_{jl}(x) \frac{\partial v_0}{\partial x_j} \frac{\partial \overline{w_0}}{\partial x_l} dx + \int_\Omega v_0 \overline{w_0} dx + \int_\Omega v_1 \overline{w_1} dx.$$

By making use of the fact that the (a_{jl}) are elliptic, i.e., satisfy condition (2.2), then it is easy to see that there exists a $C > 0$ such that

(2.99)
$$C^{-1}(||v_0||^2_{1,L^2(\Omega)} + ||v_1||^2_{L^2(\Omega)}) \leqslant (X, X)_{\mathcal{H}}$$
$$\leqslant C(||v_0||^2_{1,L^2(\Omega)} + ||v_1||^2_{L^2(\Omega)}).$$

Next, we define the operator $\mathcal{A}$. First, we set the domain $D(\mathcal{A})$ to be

(2.100)
$$D(\mathcal{A}) = \left\{ X = \{v_0, v_1\}; v_0 \in H^2(\Omega), v_1 \in H^1(\Omega), \text{ and} \right.$$
$$\left. \frac{\partial v_0}{\partial \nu_A} - \sigma_0 v_1 = 0 \text{ on } \Gamma \right\}.$$

Then, for $X = \{v_0, v_1\} \in D(\mathcal{A})$ we define

(2.101)
$$\mathcal{A}X = \{v_1, Av_0 - Hv_1\}.$$

So, $\mathcal{A}$ is a linear mapping from $D(\mathcal{A})$ to $\mathcal{H}$.

Now, using the above Hilbert space $\mathcal{H}$, the operator $\mathcal{A}$ in $\mathcal{H}$ and by letting $X(t) = \{u(t, \cdot), u_t(t, \cdot)\}$, $F(t) = \{0, f(t, \cdot)\}$, we can express (2.96) as

(2.102)
$$\begin{cases} \dfrac{dX(t)}{dt} = \mathcal{A}X(t) + F(t), \\ X(0) = \{u_0, u_1\}. \end{cases}$$

It should be noted that the boundary condition is included in $X(t) \in D(\mathcal{A})$.

PROPOSITION 2.20. *For arbitrary* $X \in D(\mathcal{A})$, *the following holds:*

(2.103)
$$\text{Re}\,(\mathcal{A}X, X)_{\mathcal{H}} \leqslant M||X||^2_{\mathcal{H}},$$

where M *is a constant independent of* X.

PROOF. By substituting (2.101) into the definition of the inner product given in (2.98), we have that

$$(\mathcal{A}X, X)_{\mathcal{H}} = \sum_{j,l=1}^{n} \int_{\Omega} a_{jl} \frac{\partial v_1}{\partial x_j} \frac{\partial \overline{v_0}}{\partial x_l} dx + \int_{\Omega} v_1 \overline{v_0} dx$$
$$+ \int_{\Omega} (Av_0 - Hv_1)\overline{v_1} dx.$$

Next, by integrating by parts, we obtain that

$$\sum_{j,l=1}^{n} \int_{\Omega} a_{jl} \frac{\partial v_1}{\partial x_j} \frac{\partial \overline{v_0}}{\partial x_l} dx = \int_{\Gamma} v_1 \frac{\partial \overline{v_0}}{\partial \nu_A} dS$$
$$+ \sum_{j,l=1}^{n} \int_{\Omega} v_1 \left(-\frac{\partial}{\partial x_j} \left(a_{jl} \frac{\partial \overline{v_0}}{\partial x_l} \right) \right) dx.$$

Also, since

$$-\operatorname{Re}(Hv_1)\overline{v_1} = -\sum_{j=1}^{n} \left\{ \frac{\partial}{\partial x_j}(h_j|v_1|^2) - \frac{\partial h_j}{\partial x_j}|v_1|^2 \right\} - \operatorname{Re}(h_0|v_1|^2),$$

we see that

$$-\operatorname{Re}\int_{\Omega}(Hv_1)\overline{v_1}dx = \int_{\Gamma} \left(-\sum_{j=1}^{n} h_j v_j \right) |v_1|^2 dS$$
$$+ \int_{\Omega} \left(\sum_{j=1}^{n} \frac{\partial h_j}{\partial x_j} - \operatorname{Re} h_0 \right) |v_1|^2 dx.$$

Therefore,

$$\operatorname{Re}(\mathcal{A}X, X)_{\mathcal{H}} = \operatorname{Re}\int_{\Gamma} v_1 \left(\overline{\frac{\partial v_0}{\partial \nu_A} - \sum_{j=1}^{n} h_j \nu_j v_1} \right) dS$$

(2.104)
$$+ \operatorname{Re}\int_{\Omega} \{v_1(-A_0\overline{v_0}) + (Av_0)\overline{v_1}\} dx$$
$$+ \int_{\Omega} \left(\sum_{j=1}^{n} \frac{\partial h_j}{\partial x_j} - \operatorname{Re} h_0 \right) |v_1|^2 dx,$$

where we set

$$A_0 v_0 = \sum_{j,l=1}^{n} \frac{\partial}{\partial x_j} \left(a_{jl} \frac{\partial v_0}{\partial x_l} \right).$$

Now, the first term in the right-hand side of (2.104) can be written as

(2.105)
$$v_1 \left(\overline{\frac{\partial v_0}{\partial \nu_A} - \sum_{j=1}^{n} h_j \nu_j v_1} \right) = v_1 \left(\overline{\frac{\partial v_0}{\partial \nu_A} - \sigma_0 v_1} \right)$$

$$+ \left(\sigma_0 - \sum_{j=1}^{n} h_j \nu_j \right) |v_1|^2.$$

Combining the fact that $X \in D(\mathcal{A})$ with the assumption, (2.11), on σ_0, we see that (2.105) must be either negative or zero. Also, we have the following estimates:

|the second term in the r.h.s. of (2.104)|

$$\leqslant \int_{\Omega} |v_1| \left| \sum_{j=1}^{n} a_j \frac{\partial v_0}{\partial x_j} + a_0 v_0 \right| dx$$

$$\leqslant ||v_1||_{L^2(\Omega)}^2 + C||v_0||_{1,L^2(\Omega)}^2,$$

|the third term in the r.h.s. of (2.104)| $\leqslant C||v_1||_{L^2(\Omega)}^2.$

From the above, we find that there exists a constant $M > 0$ which is independent of X and such that

$$\mathrm{Re}\,(\mathcal{A}X, X) \leqslant M||X||_{\mathcal{H}}^2.$$

$\square$

LEMMA 2.21. $D(\mathcal{A})$ is dense in $\mathcal{H}$.

PROOF. Suppose that $X = \{v_0, v_1\} \in \mathcal{H}$. Then with $v_{0j} \in H^\infty(\Omega)$ and $v_{1j} \in H^\infty(\Omega)$ $(j = 1, 2, \dots)$ as $j \to \infty$ the following is true:

$$v_{0j} \longrightarrow v_0 \text{ in } H^1(\Omega), \quad v_{1j} \longrightarrow v_1 \text{ in } L^2(\Omega).$$

Next, we define the function $g \in H^\infty(\Gamma)$ on Γ by

$$g_j = \left[\frac{\partial v_{0j}}{\partial \nu_A} - \sigma_0 v_{1j} \right]_\Gamma.$$

Using Lemma 2.10, we can choose $w_j \in H^\infty(\Omega)$ such that

$$\left. \frac{\partial w_j}{\partial \nu_A} \right|_\Gamma = g_j \quad \text{and} \quad ||w_j||_{1,L^2(\Omega)} \leqslant j^{-1}.$$

So, if we set $X_j = \{v_{0j} - w_j, v_{1j}\}$, we have that $X_j \in D(\mathcal{A})$ and

$$X_j \longrightarrow X \quad \text{in } \mathcal{H} \qquad (j \to \infty).$$

$\square$

In order to be able to apply Theorem 2.18 in equations (2.102), the only thing that remains to be shown is the existence of $(\lambda - \mathcal{A})^{-1}$. With this in mind, for $\lambda > M$ it is necessary to show that

$$(2.106) \qquad \lambda - \mathcal{A} : D(\mathcal{A}) \longrightarrow \mathcal{H}$$

is a bijection. From the estimate (2.103) and by the method that leads to (2.84) in the first step of Lemma 2.17, when $\lambda > M$ we have that $\lambda - \mathcal{A}$ is an injection. Hence, all that is required to be shown is that (2.106) is a surjection.

For $X = \{v_0, v_1\} \in D(\mathcal{A})$ and $F = \{f_0, f_1\} \in \mathcal{H}$

$$(2.107) \qquad (\lambda - \mathcal{A})X = F$$

is true if and only if $\lambda v_0 - v_1 = f_0$ and $-Av_0 + (\lambda + H)v_1 = f_1$.

Now, substituting $v_1 = \lambda v_0 - f_0$ in the second formula leads to

$$(2.108) \qquad -Av_0 + \lambda H v_0 + \lambda^2 v_0 = f_1 + (\lambda + H)f_0 \quad \text{in } \Omega.$$

Also, the boundary condition satisfied by $X \in D(\mathcal{A})$ can be expressed as

$$(2.109) \qquad \frac{\partial v_0}{\partial \nu_A} - \lambda \sigma_0 v_0 = \sigma_0 f_0 \qquad \text{on } \Gamma.$$

Conversely, for $\{f_0, f_1\} \in \mathcal{H}$, if there exists a $v_0 \in H^2(\Omega)$ that satisfies (2.108) and (2.109), then taking $v_1 = \lambda v_0 - f_0$ and setting $X = \{v_0, v_1\}$, we have that $X \in D(\mathcal{A})$ and, moreover, that X satisfies (2.107). From the above, the problem of finding an X that satisfies (2.107) for F is the same as the problem of finding a v_0 that satisfies (2.108) and (2.109).

So, we show the existence of a solution of (2.108) and (2.109). First, we assume that $\lambda > 0$, and for the parameter $0 \leqslant \varepsilon \leqslant \lambda$ we consider the following boundary value problem:

$$(2.110)_\varepsilon \qquad \begin{cases} L_\varepsilon v = -Av + \varepsilon H v + \lambda^2 v = f & \text{in } \Omega, \\ B_\varepsilon v = \dfrac{\partial v}{\partial \nu_A} - \varepsilon \sigma_0 v = g & \text{on } \Gamma, \end{cases}$$

in which we assume that $f \in L^2(\Omega)$ and $g \in H^{1/2}(\Gamma)$.

We would like to show that this boundary value problem, if λ is suitably large, has a solution with $0 \leqslant \varepsilon \leqslant \lambda$. Since the proof itself is quite lengthy, we will divide it into several steps. First, we prove the following lemma.

LEMMA 2.22. *Suppose that there exists a positive constant λ_2 such that $\lambda \geqslant \lambda_2$ and $0 \leqslant \varepsilon \leqslant \lambda$. If $v \in H^2(\Omega)$ is a solution of* $(2.110)_\varepsilon$, *then the following holds:*

$$(2.111) \qquad \sqrt{\frac{c_0}{2}}\|v\|_{1,L^2(\Omega)} \leqslant \|f\|_{L^2(\Omega)} + \langle g \rangle^2_{L^2(\Gamma)}.$$

PROOF. Suppose $v \in H^2(\Omega)$ is a solution. Since

$$2\text{Re} \int_\Omega (-Av)\bar{v}dx = 2\text{Re} \sum_{j,l=1}^n \int_\Omega a_{jl} \frac{\partial v}{\partial x_j} \overline{\frac{\partial v}{\partial x_l}} dx - 2\text{Re} \int_\Gamma \frac{\partial v}{\partial \nu_A} \bar{v}dS$$

$$+ 2\text{Re} \sum_{j=1}^n \int_\Omega a_j \frac{\partial v}{\partial x_j} \bar{v}dx + 2\text{Re} \int_\Omega a_0 |v|^2 dx,$$

$$2\text{Re} \int_\Omega \varepsilon(Hv)\bar{v}dx = 2\varepsilon \sum_{j=1}^n \int_\Gamma h_j \nu_j |v|^2 dS$$

$$+ 2\text{Re} \int_\Omega \varepsilon \left(h_0 - \sum_{j=1}^n \frac{\partial h_j}{\partial x_j} \right) |v|^2 dx,$$

there exist constants C_1, C_2 and C_3, independent of $0 \leqslant \varepsilon \leqslant \lambda$, such that

$$2\text{Re} \int_\Omega f\bar{v}dx \geqslant 2\text{Re} \sum_{j,l=1}^n \int a_{jl} \frac{\partial v}{\partial x_j} \overline{\frac{\partial v}{\partial x_l}} dx$$

$$+ \int_\Omega (\lambda^2 - C_1\lambda - C_2)|v|^2 dx - C_3\|v\|_{1,L^2(\Omega)}\|v\|_{L^2(\Omega)}$$

$$- 2\text{Re} \int_\Gamma \left(\frac{\partial v}{\partial \nu_A} - \varepsilon\sigma_0 v \right) \bar{v}dS + 2\varepsilon \int_\Gamma \left(\sum_{j=1}^n h_j \nu_j - \sigma_0 \right) |v|^2 dS.$$

Also, the following is true:

$$\left| \int_\Gamma B_\varepsilon v\bar{v}dx \right| = \left| \int_\Gamma g\bar{v}dx \right| \leqslant \langle g \rangle^2_{L^2(\Gamma)} + \langle v \rangle^2_{L^2(\Gamma)}.$$

So, for arbitrary $\alpha > 0$, we can estimate

$$\langle v \rangle_{L^2(\Omega)}^2 \leqslant \alpha ||v||_{1,L^2(\Omega)}^2 + C_\alpha ||v||_{L^2(\Omega)}^2.$$

Now, by using the fact that

$$C_3 ||v||_{1,L^2(\Omega)} ||v||_{L^2(\Omega)} \leqslant \alpha ||v||_{1,L^2(\Omega)}^2 + C_\alpha' ||v||_{L^2(\Omega)}^2$$

and that σ_0 satisfies condition (2.11), by taking $\alpha = \frac{1}{4}c_0$ and if necessary interchanging C_1 and C_2, we have that

$$\frac{c_0}{2} ||v||_{1,L^2(\Omega)}^2 + (\lambda^2 - C_1\lambda - C_2) ||v||_{L^2(\Omega)}^2 \leqslant ||f||_{L^2(\Omega)}^2 + \langle g \rangle_{L^2(\Gamma)}^2.$$

Finally, by choosing $\lambda_2 > 0$ in such a way that if $\lambda \geqslant \lambda_2$, then

$$\lambda^2 - C_1\lambda - C_2 \geqslant \frac{1}{2}\lambda^2 > \frac{1}{2}c_0,$$

we obtain (2.111).

$\square$

LEMMA 2.23. *Suppose that there exists a constant $\alpha > 0$ and the following holds true: Let $\lambda \geqslant \lambda_2$. Then for some $0 \leqslant \varepsilon_1 \leqslant \lambda$ the equations $(2.110)_{\varepsilon_1}$, with arbitrary $f \in L^2(\Omega)$ and $g \in H^{1/2}(\Gamma)$, have solutions in $H^2(\Omega)$.*
With the above, for an arbitrary ε that satisfies

$$|\varepsilon - \varepsilon_1| \leqslant \alpha, \qquad 0 \leqslant \varepsilon \leqslant \lambda,$$

and with arbitrary $f \in L^2(\Omega)$ and $g \in H^{1/2}(\Gamma)$, $(2.110)_\varepsilon$ has a solution in $H^2(\Omega)$.

PROOF. For arbitrary $f \in L^2(\Omega)$ and $g \in H^{1/2}(\Gamma)$, let $v^{(0)}$ be a solution of $(2.110)_{\varepsilon_1}$. From (2.111), $v^{(0)}$ has the following estimate:

$$||v^{(0)}||_{1,L(\Omega)} \leqslant \sqrt{2c_0^{-1}}\{||f||_{L^2(\Omega)} + \langle g \rangle_{L^2(\Gamma)}\}.$$

Now, suppose that $v^{(1)}$ is a solution of

$$\begin{cases} L_{\varepsilon_1}v^{(1)} = -(\varepsilon - \varepsilon_1)Hv^{(0)} & \text{in } \Omega, \\ B_{\varepsilon_1}v^{(1)} = (\varepsilon - \varepsilon_1)\sigma_0 v^{(0)} & \text{on } \Gamma. \end{cases}$$

The existence of such a $v^{(1)}$ is guaranteed by the hypothesis.

In the above, if we use (2.111), we can deduce that

$$||v^{(1)}||_{1,L(\Omega)} \leq \sqrt{2c_0^{-1}}|\varepsilon - \varepsilon_1|\{||Hv^{(0)}||_{L^2(\Omega)} + \langle\sigma_0 v^{(0)}\rangle_{L^2(\Gamma)}\}$$

$$\leq \sqrt{2c_0^{-1}}C|\varepsilon - \varepsilon_1| \times ||v^{(0)}||_{1,L^2(\Omega)}.$$

We can define $v^{(p)}$ inductively, and so let $v^{(p+1)}$ be a solution of

$$\begin{cases} L_{\varepsilon_1}v^{(p+1)} = -(\varepsilon - \varepsilon_1)Hv^{(p)} & \text{in } \Omega, \\ B_{\varepsilon_1}v^{(p+1)} = (\varepsilon - \varepsilon_1)\sigma_0 v^{(p)} & \text{on } \Gamma. \end{cases}$$

Then, $v^{(p+1)} \in H^2(\Omega)$ exists and has the following estimate:

$$||v^{(p+1)}||_{1,L^2(\Omega)} \leq \sqrt{2c_0^{-1}}C|\varepsilon - \varepsilon_1| \times ||v^{(p)}||_{1,L^2(\Omega)}.$$

Therefore, $v^{(p)}$ ($p = 0, 1, 2, \ldots$) can be defined inductively as elements of $H^2(\Omega)$, and the following estimate is true:

$$||v^{(p)}||_{1,L^2(\Omega)} \leq \left(\sqrt{2c_0^{-1}}C|\varepsilon - \varepsilon_1|\right)^p \{||f||_{L^2(\Omega)} + \langle g\rangle_{L^2(\Gamma)}\}.$$

If

(2.112)
$$\sqrt{2c_0^{-1}}C|\varepsilon - \varepsilon_1| < 1,$$

then we can easily see that as $q \to \infty$,

$$v_q = \sum_{p=0}^{q} v^{(p)}$$

converges strongly in $H^1(\Omega)$. Also, the following holds:

$$\begin{cases} L_{\varepsilon_1}v_q = -(\varepsilon - \varepsilon_1)Hv_{q-1} + f & \text{in } \Omega, \\ B_{\varepsilon_1}v_q = (\varepsilon - \varepsilon_1)\sigma_0 v_{q-1} + g & \text{on } \Gamma. \end{cases}$$

As $q \to \infty$, since the right-hand side converges, respectively, in $L^2(\Omega)$ and $H^{1/2}(\Gamma)$, then by using the *a priori* estimate of the solution, it is easy to see that v_q converges in $H^2(\Omega)$.

Now, if we set

$$v = \lim_{q\to\infty} v_q,$$

by construction we have that v satisfies

$$\begin{cases} L_{\varepsilon_1}v = -(\varepsilon - \varepsilon_1)Hv + f & \text{in } \Omega, \\ B_{\varepsilon_1}v = (\varepsilon - \varepsilon_1)\sigma_0 v + g & \text{on } \Gamma. \end{cases}$$

So, what we have shown is that v is a solution of $(2.110)_\varepsilon$. Hence, for arbitrary f and g, $(2.110)_\varepsilon$ does in fact possess a solution.

From the above, if we take $\alpha = (\sqrt{2c_0^{-1}C})^{-1}$, then we see that the lemma is true.

$\square$

Now, for the case $\varepsilon = 0$, i.e., $(2.110)_0$, for arbitrary $f \in H^2(\Omega)$ and $g \in H^{1/2}(\Gamma)$, a solution in $H^2(\Omega)$ is guaranteed by Theorem 2.9, which is the existence theorem for a solution for the boundary value problem of an elliptic equation. Even for $0 \leqslant \varepsilon \leqslant \alpha$, it is easy to see from Lemma 2.23 that there is a solution. By repeating this process, we shall eventually show that there is a solution for all ε such that $0 \leqslant \varepsilon \leqslant \lambda$. Then, if we take $\varepsilon = \lambda$, it is a straightforward matter to understand that (2.108) and (2.109) also have solutions. Hence, we have shown that $\lambda - \mathcal{A}$ is a surjection. The above work now allows us to state the following proposition.

PROPOSITION 2.24. *For $\lambda > \lambda_2$, we are assured of the existence of $(\lambda - \mathcal{A})^{-1}$.*

(c) The existence of a solution for the initial boundary value problem.

As we noted previously in Section (b), the initial boundary value problem (2.96), using the operator $\mathcal{A}$ defined by (2.100) and (2.101), can be written as

$$(2.113) \qquad \begin{cases} \dfrac{dX(t)}{dt} = \mathcal{A}X(t) + F(t), \\ X(0) = X_0, \end{cases}$$

where we set $F(t) = \{0, f(t, \cdot)\}$ and $X_0 = \{u_0, u_1\}$. The results proved in Section (b) allow us now to apply Theorem 2.18 to (2.113).

Suppose for $u_0 \in H^2(\Omega)$, $u_1 \in H^1(\Omega)$ there exists a solution u of (2.96) that belongs to $\bigcap\limits_{l=0}^{2} C^l([0, \infty); H^{2-l}(\Omega))$. Since the boundary conditions are satisfied, the following holds:

$$\frac{\partial u}{\partial \nu_A}(t, x) - \sigma_0 \frac{\partial u}{\partial t}(t, x) = 0, \qquad (t, x) \in [0, \infty) \times \Gamma.$$

If we now set $t = 0$, from the initial conditions we have that

$$(2.114) \qquad \frac{\partial u_0}{\partial \nu_A} - \sigma_0 u_1 = 0 \qquad \text{on } \Gamma.$$

That is to say, we must have that

$$(2.115) \qquad\qquad X_0 \in D(\mathcal{A}).$$

Conversely, we assume that the initial data satisfies (2.115). By assuming $f \in C([0, \infty); H^1(\Omega))$ and $f(t, x)|_\Gamma = 0$, then, from Corollary 2.19, we find that the solution $X(t)$ of (2.113) is given by

$$(2.116) \qquad X(t) = U(t)X_0 + \int_0^t U(t - s)F(s)ds.$$

If we set $X(t) = \{v_0(t), v_1(t)\}$, then, since $X(t) \in D(\mathcal{A})$, for each $t > 0$, we have that $v_0(t) \in H^2(\Omega)$ and $v_1(t) \in H^1(\Omega)$. Also, since $X'(t)$ is strongly continuous in $\mathcal{H}$, it is easy to see that

$$(2.117) \quad v_0(t) \in C^1([0, \infty); H^1(\Omega)), \quad v_1(t) \in C^1([0, \infty); L^2(\Omega)).$$

Further, by the way we have chosen $F(t)$, since $v_0'(t) = v_1(t)$, and combining this with the second formula in (2.117), we obtain that

$$(2.118) \qquad\qquad v_0(t) \in C^2([0, \infty); L^2(\Omega)).$$

If we now set $u(t, \cdot) = v_0(t)$, then, as noted at the beginning, the following is satisfied:

$$(2.119) \qquad Pu(t, x) = f(t, x) \qquad \text{in } (0, \infty) \times \Omega.$$

Also, the fact that $X(t) \in D(\mathcal{A})$, for every $t \geqslant 0$, shows that $u(t, \cdot) \in H^2(\Omega)$.

Further, by (2.119) we can write

$$Au(t, x) = -\frac{\partial^2 u}{\partial t^2}(t, x) - \left(H\frac{\partial u}{\partial t}\right)(t, x) + f(t, x) \in C^0([0, \infty); L^2(\Omega)).$$

Since the boundary conditions are also satisfied, we have that

$$\frac{\partial u}{\partial \nu_A}(t, x) = \sigma_0 \frac{\partial u}{\partial t}(t, x) \in C^0([0, \infty); H^{1/2}(\Gamma)).$$

Then, by using the *a priori* estimate for A, we get that

$$(2.120) \qquad\qquad u(t, \cdot) \in C^0([0, \infty); H^2(\Omega)).$$

So, by (2.117), (2.118) and (2.120), we can show that

$$u \in C^0([0, \infty); H^2(\Omega)) \cap C^1([0, \infty); H^1(\Omega)) \cap C^2([0, \infty); L^2(\Omega)).$$

For $f \in C^1([0, \infty); L^2(\Omega))$, we would like to note that there exists a sequence $f_j \in C^1([0, \infty); H^1(\Omega))$ such that $f_j(t, x)|_\Gamma = 0$ and $f_j \to$

f in $C^1([0,\infty); L^2(\Omega))$ as $j \to \infty$. Using this fact and Theorem 2.15, we are led to the next theorem.

THEOREM 2.25. *Suppose the initial data u_0 and u_1, with $u_0 \in H^2(\Omega)$ and $u_1 \in H^1(\Omega)$, satisfy*

$$\frac{\partial u_0}{\partial \nu_A} - \sigma_0 u_1 = 0 \qquad on \ \Gamma.$$

Also, when $f \in C^1([0,\infty); L^2(\Omega))$ suppose that (2.96) has a solution u such that

$$u \in \bigcap_{j=0}^{2} C^{2-j}([0,\infty); H^j(\Omega)).$$

Then, u satisfies the following estimate:

$$|||u(t)|||_{2,\Omega} \leqslant C \left\{ e^{Ct}(||u_0||_{2,L^2(\Omega)} + ||u_1||_{1,L^2(\Omega)} + ||f(0,\cdot)||_{L^2(\Omega)} \right.$$
$$\left. + \int_0^t e^{C(t-s)} \left|\left|\frac{\partial f}{\partial t}(s,\cdot)\right|\right|_{L^2(\Omega)} ds \right\}.$$

(d) Smoothness of the solution.

If the given initial data $\{u_0, u_1\}$ and f are smooth, the natural question to ask is: "Is the solution with this data also smooth?" To obtain an answer, we assume that $u \in \bigcap_{j=0}^{3} C^j([0,\infty); H^{3-j}(\Omega))$. By partially differentiating with respect to t the boundary condition that is satisfied by u, we obtain

(2.121) $$\frac{\partial}{\partial \nu_A}\frac{\partial u}{\partial t}(t,x) - \sigma_0(x)\frac{\partial}{\partial t}\frac{\partial u}{\partial t}(t,x) = 0 \qquad on \ \Gamma;$$

while, on the other hand,

$$\lim_{t\to+0}\frac{\partial^2 u}{\partial t^2}(t,x) = \lim_{t\to+0}\left(Au(t,x) - H\frac{\partial u}{\partial t}(t,x) + f(t,x)\right)$$
$$= (Au_0)(x) - (Hu_1)(x) + f(0,x).$$

Now, we denote the right-hand side by $u_2(x)$. Then by letting $t \to +0$ in (2.121), we can deduce that

(2.122) $$\frac{\partial}{\partial \nu_A}u_1(x) - \sigma_0 u_2(x) = 0 \qquad on \ \Gamma.$$

That is, in order for u to be in $\bigcap\limits_{j=0}^{3} C^{j}([0,\infty); H^{3-j}(\Omega))$, in addition to having $u_0 \in H^3(\Omega)$, $u_1 \in H^2(\Omega)$ and $f \in C^2(H(\Omega))$, it is also necessary that (2.122) is satisfied.

This condition can be understood as follows, the state at $t = 0$, determined by the data, and the boundary conditions must be compatible. We shall call this condition the *compatibility condition*. In passing, (2.114) is also the compatibility condition. In order to consider higher degrees of smoothness we give the next definition.

DEFINITION 2.26. For u_0, u_1 and f, we define u_p ($p = 2, 3, \ldots$) inductively by

$$(2.123) \qquad u_p = (Au_{p-2}) - (Hu_{p-1}) + \frac{\partial^{p-2} f}{\partial t^{p-2}}(0, \cdot).$$

Then the data $\{u_0, u_1, f\}$ is said to satisfy the m-order compatibility conditions if

$$u_0 \in H^{m+2}, \quad u_1 \in H^{m+1}, \quad f \in \bigcap_{j=0}^{m} C^j([0,\infty); H^{m-j}(\Omega)),$$

and for $p = 0, 1, 2, \ldots, m$, the following hold:

$$(2.124) \qquad \frac{\partial u_p}{\partial \nu_A} - \sigma_0 u_{p+1} = 0 \qquad \text{on } \Gamma.$$

NOTE. If u, the solution of (2.96), exists, u_p defined by (2.123) must be such that $u_p = (\partial_t^p u)(0, \cdot)$. (2.114) is the zero-order compatibility condition, while the data that consists of (2.114) and (2.122) satisfies the first-order compatibility condition. Further, (2.114) and $X_0 = \{u_0, u_1\} \in D(\mathcal{A})$ are equivalent. Also (2.122) and $\mathcal{A}X_0 + F(0, \cdot) \in D(\mathcal{A})$ are equivalent. Finally, for the case $f \equiv 0$, the mth-order compatibility condition and $X_0 \in D(\mathcal{A}^{m+1})$ are equivalent.

THEOREM 2.27. *Suppose the data u_0, u_1 and f satisfy the mth-order compatibility condition. Then, the solution u of the initial boundary value problem (2.96) is such that*

$$u \in \bigcap_{j=0}^{m+2} C^{m+2-j}([0,\infty); H^j(\Omega)).$$

PROOF. First, we consider the case $m = 1$, and we set $X_1 = \mathcal{A}X_0 + F(0) = \{u_1, u_2\}$. From the hypothesis, $X_1 \in D(\mathcal{A})$. Now, suppose v is a solution of

(2.125)
$$\begin{cases} Pv(t, x) = \dfrac{\partial f}{\partial t}(t, x) & \text{in } (0, \infty) \times \Omega, \\[2mm] Bv = 0 & \text{on } (0, \infty) \times \Gamma, \\[2mm] v(0, x) = u_1(x), \quad \dfrac{\partial v}{\partial t}(0, x) = u_2(x). \end{cases}$$

From Theorem 2.25, we know that the solution v exists and it belongs to $\bigcap\limits_{j=0}^{2} C^j([0, \infty); H^{2-j}(\Omega))$.

So, if we let

(2.126)
$$u(t, x) = u_0 + \int_0^t v(s, x)\,ds,$$

then

(2.127)
$$\frac{\partial}{\partial t}(Pu) = P\frac{\partial u}{\partial t} = Pv = \frac{\partial f}{\partial t}.$$

Also,

(2.128)
$$\begin{aligned} Pu|_{t=0} &= \frac{\partial v}{\partial t}(0, \cdot) + Hv(0, \cdot) - Au_0(\cdot) \\ &= u_2 + Hu_1 - Au_0 = f(0, \cdot). \end{aligned}$$

Now, (2.127) and (2.128) show that the following holds:

$$Pu(t, x) = f(t, x) \qquad \text{in } (0, \infty) \times \Omega.$$

In addition, the boundary condition is

$$\frac{\partial}{\partial t}Bu = B\frac{\partial u}{\partial t} = 0 \qquad \text{on } (0, \infty) \times \Gamma.$$

So, since

$$Bu|_{t=0} = 0 \qquad \text{on } \Gamma,$$

we obtain that

$$Bu = 0 \qquad \text{on } (0, \infty) \times \Gamma.$$

On the other hand, for $t > 0$, it is easy to see that

$$Au(t, x) = H\frac{\partial u}{\partial t}(t, x) + \frac{\partial^2 u}{\partial t^2}(t, x) - f(t, x)$$

$$= Hv(t, x) + \frac{\partial v}{\partial t}(t, x) - f(t, x) \in C^0([0, \infty); H^1(\Omega)),$$

$$\frac{\partial u}{\partial \nu_A}(t, x)\Big|_\Gamma = \sigma_0 \frac{\partial u}{\partial t}\Big|_\Gamma = \sigma_0 v|_\Gamma \in C_0^\infty([0, \infty); H^{3/2}(\Gamma)).$$

Using the estimate of the solution of the boundary value problem for the elliptic equations, we get that

$$(2.129) \qquad\qquad u \in C^0([0, \infty); H^3(\Omega)).$$

From the fact that $v \in \bigcap_{j=0}^{2} C^j([0, \infty); H^{2-j}(\Omega))$ and (2.126), we can deduce that

$$u \in C^1([0, \infty); H^2(\Omega)) \cap C^2([0, \infty); H^1(\Omega)) \cap C^3([0, \infty); L^2(\Omega)).$$

Combining this with (2.129), we find that

$$u \in \bigcap_{j=0}^{3} C^j([0, \infty); H^{3-j}(\Omega)).$$

Let us now consider the case of $m = 2$. Since $\{u_0, u_1, f\}$ satisfies the second-order compatibility condition, it follows that the data $\{u_1, u_2, \frac{\partial f}{\partial t}\}$ satisfies the first-order compatibility condition. By applying the result for $m = 1$ to (2.125), we see that

$$v \in \bigcap_{j=0}^{3} C^j([0, \infty); H^{3-j}(\Omega)).$$

If we now define u by (2.126), we can in a manner similar to that above show

$$u \in C^0([0, \infty); H^4(\Omega)).$$

Hence,

$$u \in \bigcap_{j=0}^{4} C^j([0, \infty); H^{4-j}(\Omega)).$$

The above process can now be repeated inductively for $m = 3, 4, \ldots$.

$\square$

Chapter summary.

2.1 If the phenomena governed by the second-order hyperbolic operator gives some sort of disturbance to a part, this effect is transmitted to its surroundings at finite speed. Therefore, the value of the solution at a particular time and place is determined by the data within a finite domain called the *domain of dependence*.

2.2 If the solution exists, then it necessarily satisfies a particular estimate. From this estimate, the fundamental relations of the solution depending on the data can be developed.

2.3 The initial boundary value problem can be written in the form of an evolution equation in a suitable space. The existence of the solution of the evolution equation is shown by use of the Hille-Yosida theorem.

Exercises

1. Consider the partial differential operator given by (2.1). Suppose that $\Phi(x)$ is a smooth real-valued function. Further, suppose P is transformed into $\widetilde{P}$ by the following change of variables:

$$t' = t - \Phi(x), \quad x' = x.$$

Determine an expression for $\widetilde{P}$ and the characteristic roots of $\widetilde{P}$.

2. Prove the existence of the solution of the initial value problem for the elastic equation (1.31), under the assumption that the constants α and β are positive.

3. Consider a system of first-order symmetric equations

$$L[\boldsymbol{u}](t, x) = \frac{\partial \boldsymbol{u}}{\partial t}(t, x) + \sum_{j=1}^{n} A_j(x) \frac{\partial \boldsymbol{u}}{\partial x_j}(t, x) + A_0(x)\boldsymbol{u}(t, x)$$

$$= \boldsymbol{f}(t, x),$$

where $A_j(x)$ is a smooth function, defined in $\mathbb{R}^n$, that takes its values in the $N \times N$ Hermite matrices and that belong to $\mathcal{B}$. Further, $\boldsymbol{u}$, $\boldsymbol{f}$ are $\mathbb{C}^N$-valued functions on $[0, \infty) \times \mathbb{R}^n$.

Determine the *a priori* estimate of the solution of the initial value problem given by

$$\begin{cases} L[\boldsymbol{u}](t, x) = \boldsymbol{f}(t, x) & ((t, x) \in (0, \infty) \times \mathbb{R}^n), \\ \boldsymbol{u}(0, x) = \boldsymbol{u}_0(x). \end{cases}$$

Further, investigate the existence of the solution.

4. Consider the operator in (2.75) but with friction, namely

$$Pu = \frac{\partial^2 u}{\partial t^2} - \sum_{j,k=1}^{n} \frac{\partial}{\partial x_j} \left(a_{jk}(x) \frac{\partial u}{\partial x_k} \right) + h_0(x) \frac{\partial u}{\partial t} = 0;$$

here we suppose that $h_0(x) \geqslant \alpha > 0$. Further, suppose that Ω is a finite domain that has a smooth boundary.

Show that if $u \in C^2([0, \infty) \times \Omega)$ with $u \not\equiv 0$ satisfies

$$\begin{cases} Pu = 0 & \text{in } (0, \infty) \times \Omega, \\ u = 0 & \text{on } (0, \infty) \times \Omega, \end{cases}$$

then for arbitrary $t_2 > t_1 \geqslant 0$ the following holds:

$$E(u; t_2) < E(u; t_1).$$

CHAPTER 3

The Construction of Asymptotic Solutions

For hyperbolic equations, by an explicit procedure, we will construct a function $u(t, x; k)$ with the purpose that $Pu(t, x; k)$ becomes small when we allow the parameter k in $u(t, x; k)$ to become large. This function proves to be very effective in the study of the phenomena governed by the equation $Pu = 0$. In particular, for u with large k it makes sense to consider waves with a high frequency. From the explicit construction of $u(t, x; k)$, if the frequency is high then by calling upon the laws of geometric optics we can determine with a suitable approximation how this wave will propagate itself. Using this asymptotic solution, we can study the propagation of the wave and its reflection and refraction.

3.1 An asymptotic solution of the hyperbolic equation

Recall from Chapter 2 the following form of the hyperbolic operator:

$$Pu = \frac{\partial^2 u}{\partial t^2} + \sum_{j=1}^{n} 2h_j \frac{\partial^2 u}{\partial x_j \partial t} - \sum_{j,l=1}^{n} a_{jl} \frac{\partial^2 u}{\partial x_j \partial x_l} + \sum_{j=1}^{n} a_j \frac{\partial u}{\partial x_j} + h_0 \frac{\partial u}{\partial t} + a_0 u.$$

We need to find a function $u(t, x; k)$, with parameter $k \in \mathbb{R}$, such that for a certain positive integer N the following is satisfied:

(3.1) $$Pu(t, x; k) = O(k^{-N}) \qquad (k \longrightarrow \infty).$$

In the case when k is sufficiently large, since k^{-N} is exceedingly small we have a reasonable justification in regarding Pu as 0. Therefore, since this u gives a sufficiently good approximation to the true solution for $Pu = 0$, as noted above, we may use it effectively in the study of the phenomena governed by $Pu = 0$. For $k \longrightarrow \infty$, the u that satisfies (3.1) is called an *asymptotic solution*.

With this in mind, we first consider the following form for u:

(3.2) $$u(t, x; k) = e^{ik\Phi(t,x)} v(t, x; k),$$

in which we assume that

(3.3) $$v(t, x; k) = \sum_{j=0}^{m+N} v_j(t, x)(ik)^{m-j}.$$

For the above u, we determine Pu. The first-order partial differential is

$$\frac{\partial}{\partial t} u = e^{ik\Phi}\left\{ ik\frac{\partial\Phi}{\partial t}v + \frac{\partial v}{\partial t}\right\},$$

$$\frac{\partial}{\partial x_j} u = e^{ik\Phi}\left\{ ik\frac{\partial\Phi}{\partial x_j}v + \frac{\partial v}{\partial x_j}\right\}.$$

By partially differentiating them once more, we obtain

$$\frac{\partial^2 u}{\partial t^2} = e^{ik\Phi}\left\{ (ik)^2\left(\frac{\partial\Phi}{\partial t}\right)^2 v + ik\left(2\frac{\partial\Phi}{\partial t}\frac{\partial v}{\partial t} + \frac{\partial^2\Phi}{\partial t^2}v\right) + \frac{\partial^2 v}{\partial t^2}\right\},$$

$$\frac{\partial^2 u}{\partial t \partial x_j} = e^{ik\Phi}\left\{ (ik)^2\left(\frac{\partial\Phi}{\partial t}\frac{\partial\Phi}{\partial x_j}\right)v \right.$$
$$\left. + ik\left(\frac{\partial\Phi}{\partial t}\frac{\partial v}{\partial x_j} + \frac{\partial\Phi}{\partial x_j}\frac{\partial v}{\partial t} + \frac{\partial^2\Phi}{\partial t\partial x_j}v\right) + \frac{\partial^2 v}{\partial t\partial x_j}\right\},$$

$$\frac{\partial^2 u}{\partial x_j \partial x_l} = e^{ik\Phi}\left\{ (ik)^2\left(\frac{\partial\Phi}{\partial x_j}\frac{\partial\Phi}{\partial x_l}\right)v \right.$$
$$\left. + ik\left(\frac{\partial\Phi}{\partial x_j}\frac{\partial v}{\partial x_l} + \frac{\partial\Phi}{\partial x_l}\frac{\partial v}{\partial x_j} + \frac{\partial^2\Phi}{\partial x_j\partial x_l}v\right) + \frac{\partial^2 v}{\partial x_l\partial x_j}\right\}.$$

From the above it follows that

(3.4)

$$e^{-ik\Phi}Pu = (ik)^2\left\{\left(\frac{\partial\Phi}{\partial t}\right)^2 + 2\sum_{j=1}^{n} h_j\frac{\partial\Phi}{\partial x_j}\frac{\partial\Phi}{\partial t} - \sum_{j,l=1}^{n} a_{jl}\frac{\partial\Phi}{\partial x_j}\frac{\partial\Phi}{\partial x_l}\right\}v$$

$$+ (ik)\left\{\left(2\frac{\partial\Phi}{\partial t} + \sum_{j=1}^{n} 2h_j\frac{\partial\Phi}{\partial x_j}\right)\frac{\partial v}{\partial t}\right.$$

$$+ \sum_{j=1}^{n}\left(2h_j\frac{\partial\Phi}{\partial t} - \sum_{l=1}^{n} 2a_{jl}\frac{\partial\Phi}{\partial x_l}\right)\frac{\partial v}{\partial x_j} + (P\Phi - a_0\Phi)v\bigg\}$$

$$+ Pv.$$

We now introduce the following notation:

$$\nabla \Phi = \left(\frac{\partial \Phi}{\partial t}, \frac{\partial \Phi}{\partial x_1}, \dots, \frac{\partial \Phi}{\partial x_n} \right) = (\Phi_t, \Phi_x)$$

and

(3.5) $$H(t, x, \tau, \xi) = \tau^2 + 2 \sum_{j=1}^{n} h_j(t, x) \xi_j \tau - \sum_{j,l=1}^{n} a_{jl}(t, x) \xi_j \xi_l.$$

We should point out that H is nothing more than the characteristic polynomial which up to this point we have denoted by p_0. To avoid confusion, we should say that the above H is distinct from the one given in (2.6). On occasion, to avoid cumbersome notation, we will use the simplified notation $H(\nabla \Phi)$ for $H(t, x, \Phi_t, \Phi_x)$.

Set

(3.6)

$$K = 2 \left(\frac{\partial \Phi}{\partial t} + \sum_{j=1}^{n} h_j \frac{\partial \Phi}{\partial x_j} \right) \frac{\partial}{\partial t} + 2 \sum_{j=1}^{n} \left(h_j \frac{\partial \Phi}{\partial t} - \sum_{l=1}^{n} a_{jl} \frac{\partial \Phi}{\partial x_l} \right) \frac{\partial}{\partial x_j}$$
$$+ (P\Phi - a_0 \Phi).$$

If we substitute (3.3) for v in (3.4) and then rearrange the expression in terms of powers of ik in descending order, we obtain

(3.7)
$$e^{-ik\Phi} Pu = (ik)^{m+2} H(\nabla \Phi) v_0 + (ik)^{m+1} \{ H(\nabla \Phi) v_1 + K v_0 \}$$
$$+ (ik)^m \{ H(\nabla \Phi) v_2 + K v_1 + P v_0 \}$$
$$+ (ik)^{m-1} \{ H(\nabla \Phi) v_3 + K v_2 + P v_1 \}$$
$$+ \cdots$$
$$+ (ik)^{-N+2} \{ H(\nabla \Phi) v_{m+N} + K v_{m+N-1} + P v_{m+N-2} \}$$
$$+ (ik)^{-N+1} \{ K v_{m+N} + P v_{m+N-1} \} + (ik)^{-N} P v_{m+N}.$$

So, if we can manipulate the above coefficients of the powers of ik so that they become zero, we would have reached our objective of obtaining $Pu = 0$.

With the above in mind, in the case when $H(\nabla \Phi) \neq 0$, for the coefficient of $(ik)^{m+2}$ to vanish we need to have $v_0 = 0$. Next, with $v_0 = 0$, in order for the coefficient of $(ik)^{m+1}$ to vanish we must have that $v_1 = 0$. Continuing in this fashion, we will eventually obtain that we need to have $v_j = 0$ $(j = 1, 2, \dots, m + N)$. However, even though we have achieved our objective, what we obtain, namely $u = 0$, is the trivial solution, and so we do not garner any non-trivial information on the phenomena that are governed by $Pu = 0$. Therefore, to obtain a meaningful u of the form given in (3.2) and (3.3), it is necessary that the following be satisfied:

$$(3.8) \qquad\qquad H(t, x, \nabla\Phi(t, x)) = 0.$$

In this case, if

$$(3.9) \qquad\qquad Kv_0 = 0$$

and

$$(3.10) \qquad Kv_j = -Pv_{j-1} \qquad (j = 1, 2, \ldots, m + N)$$

are satisfied, then in (3.7) besides the coefficient of $(ik)^{-N}$ all the coefficients are zero. Therefore, we see that

$$(3.11) \qquad\qquad Pu = e^{ik\Phi}(ik)^{-N}Pv_{m+N}.$$

So, this now gives us what we described earlier in this chapter; namely, for large k, u is a suitably good approximation of the true solution of $Pu = 0$. That is to say, $u(t, x; k)$ is the asymptotic solution of the equation $Pu = 0$.

Usually, the Φ and v given in (3.2) are called, respectively, the *phase function* and the *amplitude function*. As considered above, in order for (3.2) to be an asymptotic solution, it is necessary that the phase function satisfies equation (3.8) and the amplitude function v satisfies (3.9) and (3.10). Further, equation (3.8) is usually called the *Eikonal equation*, while (3.9) and (3.10) are called *transport equations*. Finally, when (3.8) is termed the *Eikonal equation*, then $H(t, x, \tau, \xi)$ is called the *Hamiltonian*.

3.2 The Eikonal equation

For (3.2) to have a non-trivial asymptotic solution, we must have that (3.8) holds.

Now, $H(t, x, \tau, \xi) = 0$ holds if and only if either $\tau = \lambda^+(t, x, \xi)$ or $\tau = \lambda^-(t, x, \xi)$, where $\lambda^\pm$ are as given in (2.4). Hence, if Φ satisfies (3.8) then either

$$(3.12) \qquad \frac{\partial\Phi}{\partial t} - \lambda^+(t, x, \Phi_x) = 0 \qquad \text{or} \qquad \frac{\partial\Phi}{\partial t} - \lambda^-(t, x, \Phi_x) = 0$$

must be satisfied. Equations of this form are called the *Hamilton-Jacobi equations*. We should note that the *Eikonal* equation is the equivalent to one or other of the Hamilton-Jacobi equations given in (3.12).

(a) Canonical equations.

Suppose that $T(s)$ and $\Theta(s)$ are real-valued functions, and let $X(s) = (X_1(s), X_2(s), \dots, X_n(s))$, $\Xi(s) = (\Xi_1(s), \Xi_2(s), \dots, \Xi_n(s))$ be functions that take their values in $\mathbb{R}^n$. Next, consider the following ordinary differential equations:

(3.13)
$$\begin{cases} \dfrac{dT(s)}{ds} = \dfrac{\partial H}{\partial \tau}(T(s), X(s), \Theta(s), \Xi(s)), \\[2mm] \dfrac{dX_j(s)}{ds} = \dfrac{\partial H}{\partial \xi_j}(T(s), X(s), \Theta(s), \Xi(s)) \qquad (j = 1, 2, \dots, n), \\[2mm] \dfrac{d\Theta(s)}{ds} = -\dfrac{\partial H}{\partial t}(T(s), X(s), \Theta(s), \Xi(s)), \\[2mm] \dfrac{d\Xi_j(s)}{ds} = -\dfrac{\partial H}{\partial x_j}(T(s), X(s), \Theta(s), \Xi(s)) \qquad (j = 1, 2, \dots, n). \end{cases}$$

The above system of ordinary differential equations is called the system of *canonical equations* for the *Eikonal* equation (3.8). Further, the solution $(T(s), X(s), \Theta(s), \Xi(s))$ of (3.13) is a curve in $\mathbb{R}^{n+1} \times \mathbb{R}^{n+1}$ usually referred to as the *bicharacteristic curve* of H, while the curve $(T(s), X(s))$ in $\mathbb{R}^{n+1}$ is called the *characteristic curve*.

Now, we suppose that the real-valued function $\Phi(t, x)$ with continuous partial derivatives of up to second order, defined in a domain of $\mathbb{R} \times \mathbb{R}^n$, satisfies the *Eikonal* equation (3.8). In addition, fix a point (t_0, x_0) in $\mathbb{R} \times \mathbb{R}^n$, and let the initial conditions for (3.13) be given by

(3.14)
$$T(0) = t_0, \ X(0) = x_0, \ \Theta(0) = \frac{\partial \Phi}{\partial t}(t_0, x_0), \ \Xi_j(0) = \frac{\partial \Phi}{\partial x_j}(t_0, x_0).$$

PROPOSITION 3.1. *In the region of s for which there exists a solution for (3.13) with the initial conditions given by (3.14), the following holds:*

(3.15) $\ H(T(s), X(s), \Theta(s), \Xi(s)) = H(t_0, x_0, \Phi_t(t_0, x_0), \Phi_x(t_0, x_0)).$

PROOF. Since

$$\frac{d}{ds}H(T(s), X(s), \Theta(s), \Xi(s))$$

$$= \frac{\partial H}{\partial t}\frac{dT}{ds} + \sum_{j=1}^{n}\frac{\partial H}{\partial x_j}\frac{dX_j}{ds} + \frac{\partial H}{\partial \tau}\frac{d\Theta}{ds} + \sum_{j=1}^{n}\frac{\partial H}{\partial \xi_j}\frac{d\Xi_j}{ds},$$

we may substitute (3.13) in the above, and it then follows that the right-hand side is zero. Hence, $H(T(s), X(s), \Theta(s), \Xi(s))$ is independent of s. Therefore, as required we obtain (3.15).   □

PROPOSITION 3.2. *In the region of s for which a solution for (3.13) with (3.14) exists and, moreover, $(T(s), X(s))$ is in the domain of Φ, we have that*

$$(3.16) \qquad \frac{\partial \Phi}{\partial t}(T(s), X(s)) = \Theta(s), \qquad \frac{\partial \Phi}{\partial x_j}(T(s), X(s)) = \Xi_j(s).$$

PROOF. First, suppose that the pair $\widetilde{T}(s)$, $\widetilde{X}(s)$ is the solution of the following system of ordinary differential equations:
(3.17)
$$\begin{cases} \dfrac{d\widetilde{T}(s)}{ds} = \dfrac{\partial H}{\partial \tau}(\widetilde{T}(s), \widetilde{X}(s), \Phi_t(\widetilde{T}(s), \widetilde{X}(s)), \Phi_x(\widetilde{T}(s), \widetilde{X}(s))), \\[2mm] \dfrac{d\widetilde{X}_j(s)}{ds} = \dfrac{\partial H}{\partial \xi_j}(\widetilde{T}(s), \widetilde{X}(s), \Phi_t(\widetilde{T}(s), \widetilde{X}(s)), \Phi_x(\widetilde{T}(s), \widetilde{X}(s))) \\[2mm] \hspace{5cm} (j = 1, 2, \dots, n), \\[2mm] \widetilde{T}(0) = t_0, \quad \widetilde{X}(0) = x_0. \end{cases}$$

Using the solutions $\widetilde{T}(s)$, $\widetilde{X}(s)$, we define $\widetilde{\Theta}(s)$ and $\widetilde{\Xi}_j(s)$ ($j = 1, 2, \dots, n$), respectively, by

$$(3.18) \qquad \widetilde{\Theta}(s) = \Phi_t(\widetilde{T}(s), \widetilde{X}(s)) \quad \text{and} \quad \widetilde{\Xi}_j(s) = \Phi_{x_j}(\widetilde{T}(s), \widetilde{X}(s)).$$

Now, by partially differentiating with respect to t both sides of (3.8), we obtain

$$\frac{\partial H}{\partial t}(t, x, \Phi_t(t, x), \Phi_x(t, x)) + \frac{\partial H}{\partial \tau}(t, x, \Phi_t(t, x), \Phi_x(t, x))\frac{\partial}{\partial t}\Phi_t(t, x)$$

$$+ \sum_{j=1}^{n}\frac{\partial H}{\partial \xi_j}(t, x, \Phi_t(t, x), \Phi_x(t, x))\frac{\partial}{\partial t}\frac{\partial \Phi}{\partial x_j}(t, x) = 0.$$

If we substitute $t = \widetilde{T}$ and $x = \widetilde{X}$ into the above, we see that

$$
\begin{aligned}
(3.19) \quad & \frac{\partial^2 \Phi}{\partial t^2}(\widetilde{T}, \widetilde{X}) \frac{\partial H}{\partial \tau}(\widetilde{T}, \widetilde{X}, \widetilde{\Phi}, \widetilde{\Xi}) \\
& + \sum_{j=1}^{n} \frac{\partial^2 \Phi}{\partial x_j \partial t}(\widetilde{T}, \widetilde{X}) \frac{\partial H}{\partial \xi_j}(\widetilde{T}, \widetilde{X}, \widetilde{\Theta}, \widetilde{\Xi}) = -\frac{\partial H}{\partial t}(\widetilde{T}, \widetilde{X}, \widetilde{\Theta}, \widetilde{\Xi}).
\end{aligned}
$$

On the other hand,

$$
\begin{aligned}
\frac{d}{ds} \widetilde{\Theta}(s) &= \frac{d}{ds}\left[\frac{\partial \Phi}{\partial t}(\widetilde{T}, \widetilde{X})\right] \\
&= \frac{\partial}{\partial t}\frac{\partial \Phi}{\partial t}(\widetilde{T}, \widetilde{X}) \frac{d\widetilde{T}}{ds} + \sum_{j=1}^{n} \frac{\partial}{\partial x_j}\frac{\partial \Phi}{\partial t}(\widetilde{T}, \widetilde{X}) \frac{d\widetilde{X}_j}{ds} \\
&= \frac{\partial^2 \Phi}{\partial t^2}(\widetilde{T}, \widetilde{X}) \frac{\partial H}{\partial \tau}(\widetilde{T}, \widetilde{X}, \widetilde{\Theta}, \widetilde{\Xi}) \\
&\quad + \sum_{j=1}^{n} \frac{\partial^2 \Phi}{\partial x_j \partial t}(\widetilde{T}, \widetilde{X}) \frac{\partial H}{\partial \xi_j}(\widetilde{T}, \widetilde{X}, \widetilde{\Theta}, \widetilde{\Xi}).
\end{aligned}
$$

In the right-hand side of the above, we can use (3.19), and this gives us that

$$
(3.20) \qquad \frac{d}{ds} \widetilde{\Theta}(s) = -\frac{\partial H}{\partial t}(\widetilde{T}(s), \widetilde{X}(s), \widetilde{\Theta}(s), \widetilde{\Xi}(s)).
$$

We can deal with the case $\widetilde{\Xi}_j(s) = \dfrac{\partial \Phi}{\partial x_j}(\widetilde{T}, \widetilde{X})$ by similar reasoning.

First, we differentiate partially with respect to x_j both sides of (3.8); then this leads to
(3.21)
$$
\begin{aligned}
& \frac{\partial H}{\partial x_j}(t, x, \Phi_t(t, x), \Phi_x(t, x)) + \frac{\partial H}{\partial \tau}(t, x, \Phi_t(t, x), \Phi_x(t, x)) \frac{\partial}{\partial x_j}\frac{\partial \Phi}{\partial t}(t, x) \\
& + \sum_{l=1}^{n} \frac{\partial H}{\partial \xi_l}(t, x, \Phi_t(t, x), \Phi_x(t, x)) \frac{\partial}{\partial x_j}\frac{\partial \Phi}{\partial x_l}(t, x) = 0.
\end{aligned}
$$

Also,

$$
\begin{aligned}
\frac{d}{ds}\widetilde{\Xi}_j(s) &= \frac{d}{ds}\left[\frac{\partial\Phi}{\partial x_j}(\widetilde{T}(s),\widetilde{X}(s))\right] \\
&= \frac{\partial}{\partial t}\frac{\partial\Phi}{\partial x_j}(\widetilde{T},\widetilde{X})\frac{d\widetilde{T}}{ds} + \sum_{l=1}^{n}\frac{\partial}{\partial x_l}\frac{\partial\Phi}{\partial x_j}(\widetilde{T},\widetilde{X})\frac{d\widetilde{X}_l}{ds} \\
&= \frac{\partial^2\Phi}{\partial t\partial x_j}(\widetilde{T},\widetilde{X})\frac{\partial H}{\partial\tau}(\widetilde{T},\widetilde{X},\widetilde{\Theta},\widetilde{\Xi}) \\
&\quad + \sum_{l=1}^{n}\frac{\partial^2\Phi}{\partial x_l\partial x_j}(\widetilde{T},\widetilde{X})\frac{\partial H}{\partial\xi_l}(\widetilde{T},\widetilde{X},\widetilde{\Theta},\widetilde{\Xi}).
\end{aligned}
$$

By using the equation in (3.21) with the substitution $t = \widetilde{T}$ and $x = \widetilde{X}$ in the right-hand side of the above, we obtain that

$$
(3.22) \qquad \frac{d}{ds}\widetilde{\Xi}_j(s) = -\frac{\partial H}{\partial\xi_j}(\widetilde{T},\widetilde{X},\widetilde{\Theta},\widetilde{\Xi}).
$$

Gathering together (3.17), (3.20) and (3.22), we see that $\widetilde{T}(s)$, $\widetilde{X}(s),\widetilde{\Theta}(s),\widetilde{\Xi}(s)$ is a solution of the system of ordinary differential equations in (3.13). Further, it is also true that $\widetilde{T},\widetilde{X},\widetilde{\Theta},\widetilde{\Xi}$ at $s = 0$ satisfy (3.14). Therefore, due to the uniqueness of the solution of the initial value problem of the ordinary differential equation in (3.13), we must have that

$$
\widetilde{T}(s) \equiv T(s), \ \ \widetilde{X}(s) \equiv X(s), \ \ \widetilde{\Theta}(s) \equiv \Theta(s), \ \ \widetilde{\Xi}(s) \equiv \Xi(s).
$$

So, by the above equalities and the definitions of Θ and Ξ given in (3.18), we obtain the desired result; namely, (3.16).   □

From the above result, we can gain an understanding of the motion of $\Phi(T, X)$. First, note that since $H(t, x, \lambda\tau, \lambda\xi) = \lambda^2 H(t, x, \tau, \xi)$ ($\lambda > 0$), we have

$$
(3.23) \qquad \tau\frac{\partial H}{\partial\tau}(t,x,\tau,\xi) + \sum_{j=1}^{n}\xi_j\frac{\partial H}{\partial\xi_j}(t,x,\tau,\xi) = 2H(t,x,\tau,\xi).
$$

Next, if the solution of (3.13) with (3.14) is T, X, Θ, Ξ, then

$$\frac{d}{ds}\Phi(T,X) = \frac{\partial\Phi}{\partial t}(T,X)\frac{dT}{ds} + \sum_{j=1}^{n}\frac{\partial\Phi}{\partial x_j}(T,X)\frac{dX_j}{ds}$$

$$= \Theta\frac{\partial H}{\partial\tau}(T,X,\Theta,\Xi) + \sum_{j=1}^{n}\Xi_j\frac{\partial H}{\partial\xi_j}(T,X,\Theta,\Xi)$$

$$= 2H(T,X,\Theta,\Xi) = 2H(t_0,x_0,\Phi_t(t_0,x_0),\Phi_x(t_0,x_0)),$$

with the last equality coming from Proposition 3.1. From the above, we can write down the next proposition.

PROPOSITION 3.3. *If $\Phi(t,x)$ satisfies* (3.8), *then for T,X of the solution of* (3.13) *with* (3.14) *we have that*

$$(3.24) \qquad \Phi(T(s),X(s)) = \Phi(t_0,x_0).$$

(b) The construction of the solution.

Following an approach opposite to that taken thus far, we look at the construction of the solution Φ for (3.8). Proposition 3.3 shows that if we assume that Φ is a solution of (3.8), then the value of Φ is fixed on the curve $\{(T(s),X(s)); s \in I\}$ (the interval I contains 0) which is determined from the solution of (3.13) with (3.14). Now, we fix t_0 and x_0. Next, for $\alpha > 0$ take a neighbourhood $U = \{x; |x - x_0| < \alpha\}$ of x_0 in $\mathbb{R}^n$ and a real-valued function $\varphi \in C^1(U)$. Then, we assume there exists a solution $\Phi(t,x)$ of (3.8) such that

$$(3.25) \qquad \Phi(t_0,x) = \varphi(x) \qquad \forall x \in U.$$

Since $\Phi_x(t_0,x) = \varphi_x(x)$, we should expect the following to hold:

$$(3.26) \qquad H(t_0,x,\Phi_t(t_0,x),\varphi_x(x)) = 0, \qquad \forall x \in U.$$

Further, suppose that the real-valued function $\theta(x) \in C(U)$ satisfies

$$(3.27) \qquad H(t_0,x,\theta(x),\varphi_x(x)) = 0, \qquad \forall x \in U.$$

Take $(T(s,y),X(s,y),\Theta(s,y),\Xi(s,y))$ to be a solution of the differential equations, (3.13), when we replace the initial conditions (3.14) with

$$(3.28) \qquad T(0) = t_0, \quad X(0) = y, \quad \Theta(0) = \theta(y), \quad \Xi(0) = \varphi_x(y),$$

where $y \in U$.

If necessary we can replace U by something smaller so there exists an open interval I, containing 0, such that for all $y \in U$ the solution that satisfies (3.13) and (3.28) exists in I.

Next, we consider the following mapping:

$$I \times U \ni (s, y) \longrightarrow (T(s, y), X(s, y)) \in \mathbb{R}^{n+1}.$$

The Jacobian of this mapping is

$$
(3.29) \qquad J(s, y) = \frac{D(T, X)}{D(s, y)} = \det
\begin{bmatrix}
\frac{\partial T}{\partial s} & \frac{\partial X_1}{\partial s} & \cdots & \frac{\partial X_n}{\partial s} \\
\frac{\partial T}{\partial y_1} & \frac{\partial X_1}{\partial y_1} & \cdots & \frac{\partial X_n}{\partial y_1} \\
\vdots & \vdots & & \vdots \\
\frac{\partial T}{\partial y_n} & \frac{\partial X_1}{\partial y_n} & \cdots & \frac{\partial X_n}{\partial y_n}
\end{bmatrix},
$$

and since $X(0, y) = y$ and $T(0, y) = t_0$, it follows that $J(0, y) = \frac{\partial T}{\partial s}(0, y)$.

On the other hand, from (3.13) we have that

$$
(3.30) \qquad \frac{\partial T}{\partial s}(0, y) = \frac{\partial H}{\partial \tau}(t_0, y, \theta(y), \varphi_x(y)).
$$

In the above, let us assume that

$$
(3.31) \qquad \frac{\partial H}{\partial \tau}(t_0, x_0, \tau_0, \varphi_x(x_0)) \neq 0,
$$

in which we set $\tau_0 = \theta(x_0)$.

Since $J(0, x_0) \neq 0$ and by applying the inverse function theorem, we see that there exist an interval $\widetilde{I}$ that contains 0, a neighbourhood $\widetilde{U}$ of x_0 in $\mathbb{R}^n$ and a neighbourhood $\widetilde{V}$ of (t_0, x_0) in $\mathbb{R}^{n+1}$, and further

$$\widetilde{I} \times \widetilde{U} \ni (s, y) \longrightarrow (T(s, y), X(s, y)) \in \widetilde{V}$$

is a bijection.

Therefore, for arbitrary $(t, x) \in \widetilde{V}$, there exists $(s, y) \in \widetilde{I} \times \widetilde{U}$ such that $(t, x) = (T(s, y), X(s, y))$. Now, using Proposition 3.3, if the solution $\Phi(t, x)$ of the *Eikonal* equation (3.8) exists, then it is easy to see that the following must be true:

$$
(3.32) \qquad \Phi(t, x) = \varphi(y).
$$

As before, we assume the existence of a $\theta(x)$ that satisfies (3.27), but here τ_0 is taken such that

$$
(3.33) \qquad H(x_0, t_0, \tau_0, \varphi_x(x_0)) = 0.
$$

If we assume (3.31), then by the use of the implicit function theorem in a sufficiently small neighbourhood of x_0, we get that $\theta(x)$, with $\theta(x_0) = \tau_0$ and satisfying (3.27), is uniquely determined.

Collecting all of the above together leads us to the following proposition.

PROPOSITION 3.4. *Suppose that φ is a function that is defined in a neighbourhood of x_0 and with continuous partial derivatives of first order. Further, suppose that τ_0 satisfies both (3.31) and (3.33). Under these circumstances, we can take a continuous function $\theta(x)$ to be defined in a neighbourhood U of x_0 that satisfies $\theta(x_0) = \tau_0$ and (3.27).*

Also, there is a neighbourhood V of (t_0, x_0) in $\mathbb{R}^{n+1}$ such that for arbitrary $(t, x) \in V$, $(s, y) \in I \times U$ uniquely exists, and using the solution of (3.13) with the initial condition (3.28) we can write

$$(3.34) \qquad (t, x) = (T(s, y), X(s, y)).$$

Finally, if there is a solution of (3.8) that satisfies (3.25) and has $\Phi_t(t_0, x_0) = \tau_0$, then the value of $\Phi(t, x)$ is determined uniquely by (3.32).

Next, we investigate whether $\Phi(t, x)$, defined by (3.32) and given in a neighbourhood of (t_0, x_0), becomes a solution of (3.8).

For the solution $(T(s, y), X(s, y), \Theta(s, y), \Theta(s, y))$ of (3.13) and (3.28), we can appropriate the proof of Proposition 3.1, and so it can be seen that the following holds:

$$H(T(s, y), X(s, y), \Theta(s, y), \Xi(s, y)) = H(t_0, y, \theta(y), \varphi_x(y)) = 0.$$

If we partially differentiate with respect to y_j the left-hand side of the above, we obtain that

$$\frac{\partial H}{\partial t}(T, X, \Theta, \Xi)\frac{\partial T}{\partial y_j} + \sum_{l=1}^{n} \frac{\partial H}{\partial x_l}(T, X, \Theta, \Xi)\frac{\partial X_l}{\partial y_j}$$

$$+ \frac{\partial H}{\partial \tau}(T, X, \Theta, \Xi)\frac{\partial \Theta}{\partial y_j} + \sum_{l=1}^{n} \frac{\partial H}{\partial \xi_l}(T, X, \Theta, \Xi)\frac{\partial \Xi_l}{\partial y_j} = 0.$$

Using (3.13) in the above equation, we derive the expression

$$(3.35) \qquad -\frac{\partial \Theta}{\partial s}\frac{\partial T}{\partial y_j} - \sum_{l=1}^{n} \frac{\partial \Xi_l}{\partial s}\frac{\partial X_l}{\partial y_j} + \frac{\partial T}{\partial s}\frac{\partial \Theta}{\partial y_j} + \sum_{l=1}^{n} \frac{\partial X_l}{\partial s}\frac{\partial \Xi_l}{\partial y_j} = 0.$$

Now,

(3.36)
$$\frac{\partial}{\partial s}\left(\Theta\frac{\partial T}{\partial y_j} + \sum_{l=1}^{n}\Xi_l\frac{\partial X_l}{\partial y_j}\right)$$
$$= \frac{\partial\Theta}{\partial s}\frac{\partial T}{\partial y_j} + \sum_{l=1}^{n}\frac{\partial\Xi_l}{\partial s}\frac{\partial X_l}{\partial y_j} + \Theta\frac{\partial}{\partial s}\frac{\partial T}{\partial y_j} + \sum_{l=1}^{n}\Xi_l\frac{\partial}{\partial s}\frac{\partial X_l}{\partial y_j},$$

and using (3.35) to replace the first half of the right-hand side, then exchanging the order of the differentiation in the latter half, we obtain

$$= \frac{\partial T}{\partial s}\frac{\partial\Theta}{\partial y_j} + \sum_{l=1}^{n}\frac{\partial X_l}{\partial s}\frac{\partial\Xi_l}{\partial y_j} + \Theta\frac{\partial}{\partial y_j}\frac{\partial T}{\partial s} + \sum_{l=1}^{n}\Xi_l\frac{\partial}{\partial y_j}\frac{\partial X_l}{\partial s}$$
$$= \frac{\partial}{\partial y_j}\left(\Theta\frac{\partial T}{\partial s} + \sum_{l=1}^{n}\Xi_l\frac{\partial X_l}{\partial s}\right).$$

Again using (3.13), this finally leads us to

(3.37)
$$\Theta\frac{\partial T}{\partial s} + \sum_{l=1}^{n}\Xi_l\frac{\partial X_l}{\partial s} = \Theta\frac{\partial H}{\partial \tau}(T, X, \Theta, \Xi) + \sum_{l=1}^{n}\Xi_l\frac{\partial H}{\partial \xi_l}(T, X, \Theta, \Xi)$$
$$= 2H(T, X, \Theta, \Xi) = 0.$$

This now allows us to say that the right-hand side of (3.36) is zero. Hence, for $j = 1, 2, \ldots, n$, it can be shown that the following holds:

(3.38)
$$\Theta(s, y)\frac{\partial T(s, y)}{\partial y_j} + \sum_{l=1}^{n}\Xi_l(s, y)\frac{\partial X_l(s, y)}{\partial y_j}$$
$$= \Theta(0, y)\frac{\partial T(0, y)}{\partial y_j} + \sum_{l=1}^{n}\Xi_l(0, y)\frac{\partial X_l(0, y)}{\partial y_j}$$
$$= \theta(y)\cdot 0 + \sum_{l=1}^{n}\frac{\partial\varphi(y)}{\partial y_l}\delta_{jl} = \frac{\partial\varphi(y)}{\partial y_j}.$$

In fact, we can write (3.37) and (3.38) in the form of a matrix; i.e.,

$$
(3.39) \quad
\begin{bmatrix}
0 \\
\dfrac{\partial \varphi}{\partial y_1} \\
\vdots \\
\dfrac{\partial \varphi}{\partial y_n}
\end{bmatrix}
=
\begin{bmatrix}
\dfrac{\partial T}{\partial s} & \dfrac{\partial X_1}{\partial s} & \cdots & \dfrac{\partial X_n}{\partial s} \\
\dfrac{\partial T}{\partial y_1} & \dfrac{\partial X_1}{\partial y_1} & \cdots & \dfrac{\partial X_n}{\partial y_1} \\
\vdots & \vdots & & \vdots \\
\dfrac{\partial T}{\partial y_n} & \dfrac{\partial X_1}{\partial y_n} & \cdots & \dfrac{\partial X_n}{\partial y_n}
\end{bmatrix}
\begin{bmatrix}
\Theta \\
\Xi_1 \\
\vdots \\
\Xi_n
\end{bmatrix} .
$$

With the hypothesis of (3.31), we define the function Φ on V by (3.32). Suppose, now, that the correspondence between (t, x) and y is given by (3.34). So, we have that

$$
(3.40) \quad
\begin{cases}
\dfrac{\partial \Phi}{\partial t}(t, x) = \displaystyle\sum_{l=1}^{n} \dfrac{\partial \varphi}{\partial y_l}(y)\dfrac{\partial y_l}{\partial t}, \\[4mm]
\dfrac{\partial \Phi}{\partial x_j}(t, x) = \displaystyle\sum_{l=1}^{n} \dfrac{\partial \varphi}{\partial y_l}(y)\dfrac{\partial y_l}{\partial x_j}.
\end{cases}
$$

Now, the Jacobian matrix of the change of variable $(s, y) \to (t, x)$ given by (3.34) is a square matrix whose determinant is the right-hand side of (3.29), and is also expressed by the right-hand side of (3.39). We should note that $I \times U$ is chosen so that the inverse matrix does exist. In this case, the Jacobian matrix of the opposite change of variable $(t, x) \to (s, y)$ is given as the inverse matrix of the previous matrix. That is,

$$
\begin{bmatrix}
\dfrac{\partial s}{\partial t} & \dfrac{\partial y_1}{\partial t} & \cdots & \dfrac{\partial y_n}{\partial t} \\
\dfrac{\partial s}{\partial x_1} & \dfrac{\partial y_1}{\partial x_1} & \cdots & \dfrac{\partial y_n}{\partial x_1} \\
\vdots & \vdots & & \vdots \\
\dfrac{\partial s}{\partial x_n} & \dfrac{\partial y_1}{\partial x_n} & \cdots & \dfrac{\partial y_n}{\partial x_n}
\end{bmatrix}
=
\begin{bmatrix}
\dfrac{\partial T}{\partial s} & \dfrac{\partial X_1}{\partial s} & \cdots & \dfrac{\partial X_n}{\partial s} \\
\dfrac{\partial T}{\partial y_1} & \dfrac{\partial X_1}{\partial y_1} & \cdots & \dfrac{\partial X_n}{\partial y_1} \\
\vdots & \vdots & & \vdots \\
\dfrac{\partial T}{\partial y_n} & \dfrac{\partial X_1}{\partial y_n} & \cdots & \dfrac{\partial X_n}{\partial y_n}
\end{bmatrix}^{-1} .
$$

If we multiply, from the left, both sides of (3.39) by the above matrices, we obtain that

$$
\begin{bmatrix} \Theta \\ \Xi_1 \\ \vdots \\ \Xi_n \end{bmatrix} = \begin{bmatrix} \dfrac{\partial s}{\partial t} & \dfrac{\partial y_1}{\partial t} & \cdots & \dfrac{\partial y_n}{\partial t} \\ \dfrac{\partial s}{\partial x_1} & \dfrac{\partial y_1}{\partial x_1} & \cdots & \dfrac{\partial y_n}{\partial x_1} \\ \vdots & \vdots & & \vdots \\ \dfrac{\partial s}{\partial x_n} & \dfrac{\partial y_1}{\partial x_n} & \cdots & \dfrac{\partial y_n}{\partial x_n} \end{bmatrix} \begin{bmatrix} 0 \\ \dfrac{\partial \varphi}{\partial y_1} \\ \vdots \\ \dfrac{\partial \varphi}{\partial y_n} \end{bmatrix}.
$$

From these relations and (3.40), we see that

$$
\Theta(s, y) = \sum_{l=1}^{n} \frac{\partial \varphi}{\partial y_l}(y) \frac{\partial y_l}{\partial t} = \frac{\partial \Phi}{\partial t}(t, x),
$$

$$
\Xi_j(s, y) = \sum_{l=1}^{n} \frac{\partial \varphi}{\partial y_l}(y) \frac{\partial y_l}{\partial x_j} = \frac{\partial \Phi}{\partial x_j}(t, x) \quad (j = 1, 2, \ldots, n).
$$

Hence,

$$
H(t, x, \Phi_t(t, x), \Phi_x(t, x)) = H(T(s, y), X(s, y), \Theta(s, y), \Xi(s, y)) = 0,
$$

and it is easy to see that $\Phi(t, x)$ becomes a solution of (3.8).

The above now yields the next theorem.

THEOREM 3.5. *Suppose that $\varphi(x)$ is a real-valued function defined in a neighbourhood of x_0 with continuous partial derivatives of first order. If both (3.31) and (3.33) hold at $\tau_0 \in \mathbb{R}$, then there exists a neighbourhood V of (t_0, x_0) and a real-valued function $\Phi(t, x)$ defined on V with continuous partial derivatives of first order that satisfies*

$$
\begin{cases} H(t, x, \Phi_t(t, x), \Phi_x(t, x)) = 0, \quad \forall (t, x)) \in V, \\ \Phi(t_0, x) = \varphi(x), \quad x \in V \cap \{t = t_0\}, \\ \Phi_t(t_0, x_0) = \tau_0. \end{cases}
$$

If $(T(s, y), X(s, y), \Theta(s, y), \Xi(s, y))$ is a solution of (3.13) with the initial condition (3.28), then the following relations hold:

(3.41)
$$
\begin{cases} \Phi(T(s, y), X(s, y)) = \varphi(s), \\ \Phi_t(T(s, y), X(s, y)) = \Theta(s, y), \\ \Phi_{x_j}(T(s, y), X(s, y)) = \Xi_j(s, y) \quad (j = 1, 2, \ldots, n). \end{cases}
$$

3.3 The transport equation and the behaviour of the asymptotic solution

(a) The transport equation.

Suppose that $\Phi(t, x)$ is a smooth function satisfying the *Eikonal* equation, (3.8). For (3.11) to hold it is necessary that both (3.9) and (3.10) are true. So, we look carefully at the first-order differentiable operator K given by (3.6). In this regard, we would like to consider first the following equation:

$$(3.42) \qquad Kw(t, x) = h(t, x).$$

Next, we assume that $T(s, y)$ and $X(s, y)$ are determined from the solution of (3.13) and (3.28).

Now, using (3.13) and (3.5), then applying (3.41), we see that

$$(3.43)$$
$$\frac{\partial T}{\partial s} = \frac{\partial H}{\partial \tau}(T, X, \Theta, \Xi) = 2\left(\Theta + \sum_{l=1}^{n} h_l(T, X)\Xi_l\right)$$
$$= 2\left(\Phi_t(T, X) + \sum_{l=1}^{n} h_l(T, X)\Phi_{x_l}(T, X)\right).$$

Similarly, we have that

$$(3.44) \qquad \frac{\partial X_j}{\partial s} = 2\left(h_j(T, X)\Phi_t(T, X) - \sum_{l=1}^{n} a_{jl}(T, X)\Phi_{x_l}(T, X)\right).$$

Therefore, if we set

$$(3.45) \qquad w(T(s, y), X(s, y)) = \widetilde{w}(s, y),$$

we get

$$
\frac{\partial \widetilde{w}}{\partial s}(s,y)
$$

$$
= \frac{\partial w}{\partial t}(T,X)\frac{\partial T}{\partial s} + \sum_{l=1}^{n}\frac{\partial w}{\partial x_j}(T,X)\frac{\partial X_j}{\partial s}
$$

$$
= 2\left\{ \Phi_t(T,X) + \sum_{l=1}^{n} h_l(T,X)\Phi_{x_l}(T,X) \right\}\frac{\partial w}{\partial t}(T,X)
$$

$$
+ 2\sum_{j=1}^{n}\left\{ h_j(T,X)\Phi_t(T,X) - \sum_{l=1}^{n} a_{jl}(T,X)\Phi_{x_l}(T,X) \right\}
$$

$$
\times \frac{\partial w}{\partial x_j}(T,X).
$$

That is to say, the following relationship holds:

(3.46)
$$
\frac{\partial \widetilde{w}}{\partial s}(s,y) = [Kw - (P\Phi - a_0\Phi)w]_{t=T,x=X}
$$
$$
= \widetilde{h}(s,y) - \widetilde{\kappa}(s,y)\widetilde{w}(s,y),
$$

where $\widetilde{\kappa}(s,y) = [P\Phi - a_0\Phi]_{t=T,x=X}$.

The important thing to realize about the above is that if we fix y, then (3.46) is an ordinary differential equation in s. In this regard, we can express the solution of the ordinary differential equation (3.46) in s as

(3.47)
$$
\widetilde{w}(s,y) = \widetilde{w}(0,y)\exp\left(-\int_0^s \widetilde{\kappa}(l,y)dl\right)
$$
$$
+ \int_0^s \widetilde{h}(\tau,y)\exp\left(-\int_\tau^s \widetilde{\kappa}(l,y)dl\right)d\tau.
$$

Conversely, suppose $w_0(x)$ is a given function. Then we would like to consider the existence of a solution of equation (3.42) that satisfies the initial condition,

(3.48)
$$
w(t_0,x) = w_0(x).
$$

From our previous deliberations, if a solution does exist, then we know that besides $\widetilde{w}(s,y)$, obtained by setting $\widetilde{w}(0,y) = w_0(y)$ in (3.47), there can be no other solution. Hence, we need to show that in fact the solution is $\widetilde{w}$ and is given by (3.47). Let $I \times U$ and a neighbourhood V of (t_0, x_0) be as in the proof of Theorem 3.5. Now, $\widetilde{w}$ given by (3.47) satisfies

$$\frac{\partial \widetilde{w}}{\partial s} + \widetilde{\kappa}(s, y)\widetilde{w} = \widetilde{h}.$$

Define $w(t, x)$ by $w(T(s, y), X(s, y)) = \widetilde{w}(s, y)$, and so

$$\frac{\partial \widetilde{w}}{\partial s} = \frac{\partial w}{\partial t}\frac{\partial T}{\partial s} + \sum_{j=1}^{n} \frac{\partial w}{\partial x_j}\frac{\partial X_j}{\partial s},$$

but since T and X are a solution of (3.13), (3.43) and (3.44) hold.

Therefore, from the definition of $\widetilde{\kappa}$, it is easy to see that w becomes a solution of (3.42). In addition, from $w(t_0, x) = \widetilde{w}(0, x) = w_0(x)$, it also follows that (3.48) holds.

PROPOSITION 3.6. *Assume that the neighbourhood V of (t_0, x_0) is chosen as in Theorem 3.5. Further, let $w_0(x)$ be a C^1-function on $V \cap \{t = t_0\}$ and let h be a continuous function on V. Then, the solution of (3.42) that satisfies the initial condition (3.48) exists and, moreover, the solution is unique.*

Using this expression, we try to construct v_j $(j = 0, 1, 2, \ldots, m + N)$ so that they satisfy (3.9) and (3.10). With this in mind, let U be a neighbourhood of x_0, and suppose $v(t_0, x; k)$ is given and satisfies

(3.49) $\operatorname{supp} v_j(t_0, \cdot) \subset U$ $(j = 0, 1, 2, \ldots, m + N)$.

Since $X(0, y) = y$, for $\widetilde{v}_j(s, y) = v_j(T(s, y), X(s, y))$,

(3.50) $\operatorname{supp} \widetilde{v}_j(0, \cdot) \subset U$.

Therefore, if we apply (3.47) with $\widetilde{h} \equiv 0$, we get

(3.51) $\widetilde{v}_0(s, y) = v_0(t_0, y)\exp\left(-\int_0^s \widetilde{\kappa}(\sigma, y)d\sigma\right).$

So, for all s,

(3.52) $\operatorname{supp} \widetilde{v}_0(s, \cdot) \subset U$.

From this, by denoting s by $s(t, y)$ with $t = T(s, y)$, we have that

(3.53) $\operatorname{supp} v_0(t, \cdot) \subset \{X(s(t, y), y); y \in U\}.$

Next in our line of considerations is v_1. If we set $Pv_0(T, X) = \widetilde{h}_0(s, y)$, then from (3.52) we see that

$$\operatorname{supp} \widetilde{h}(s, \cdot) \subset U \quad \forall s \in I.$$

Now, by applying (3.47) to $\widetilde{v}_1$, we obtain

$$\operatorname{supp} \widetilde{v}_1(s, \cdot) \subset U \quad \forall s \in I.$$

Continuing in the above fashion, we eventually establish that for $j = 0, 1, 2, \ldots, m + N$,

$$\operatorname{supp} \widetilde{v}_j(s, \cdot) \subset U \quad \forall s \in I.$$

The above now yields the following proposition.

PROPOSITION 3.7. *For* $v(t_0, x; k)$, *we assume* (3.49). *Then, for the support of* $v_j(t, x)$ $(j = 0, 1, 2, \ldots, m + N)$ *that satisfies* (3.9) *and* (3.10), *the following holds:*

$$\operatorname{supp} v_j(t, \cdot) \subset \{X(s, y); T(s, y) = t, \ y \in U\}.$$

(b) The behaviour of the asymptotic solution.
When k is sufficiently large, the principal part of u in (3.2) and (3.3) is

$$u_0(t, x; k) = e^{ik\Phi(t,x)} v_0(t, x).$$

So, we suppose that $t_0 = 0$ and $\Phi(0, x) = \varphi(x)$. Further, let $u_0(T, X; k) = \widetilde{u}_0(s, y; k)$; then from (3.32) and (3.51) we have that

$$(3.54) \qquad \widetilde{u}_0(s, y; k) = e^{ik\varphi(y)} v_0(0, y) \exp\left(-\int_0^s \widetilde{\kappa}(l, y) dl\right).$$

Although the behaviour of u_0 along the characteristic curve does change considerably, the topology itself does not change.

Now consider the behaviour of u_0 based on (3.54). In order to simplify matters, we assume that H is independent of the variable t and $h_j(x) \equiv 0$ $(j = 1, 2, \ldots, n)$. As a result of these assumptions, in the canonical equations (3.13) we get $\dfrac{d\Theta}{ds} = 0$, and hence $\Theta(s)$ is a constant. Also,

$$\frac{dT(s)}{ds} = 2\Theta(s),$$

and if we suppose that φ is given in such a way that $\Phi_t(0, x) = 1/2$, then in the range, where Φ exists, we see that $\Theta(s) = 1/2$. Therefore, $t = s$.

Now, if we assume that U in (3.49) is a sufficiently small neighbourhood of y_0, then $X(t, y)$ may be regarded as the linear approximation

$$X(t, y) \fallingdotseq X(t, y_0) + R(t)(y - y_0),$$

where $R(t)$ is the $n \times n$ matrix given by $\left. \dfrac{\partial X(t, y)}{\partial y} \right|_{y=y_0}$, and $y - y_0$ is deemed to be the column vector. So, if we set $X(t, y) = x$, then

$$y = y_0 + R(t)^{-1}(x - X(t, y_0)).$$

Next, if we substitute this relation into (3.54), we see that
(3.55)
$$u_0(t, x; k) = u_0(0, y_0 + R(t)^{-1}(x - X(t, y_0)); k) \exp\left(-\int_0^t \widetilde{\kappa}(l, y)dl\right).$$

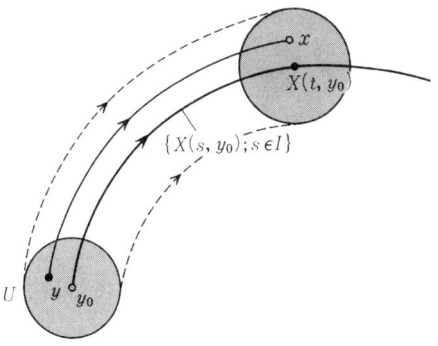

FIGURE 3.1

This formula shows that with time the support of $u_0(t, x)$, as a function x, moves along the characteristic curve $\{X(s, y_0); s \geqslant 0\}$ with starting point y_0. So, for each fixed t, the graph as a function of x is as follows. First, $u_0(0, \cdot; k)$ is translated in a parallel manner up to $X(t, y_0)$. Next, it expands or contracts and rotates by a factor determined by $R(t)$. Finally, it changes size by $\exp\left(-\int_0^t \widetilde{\kappa}(l, y)dl\right)$.

Next, let us fix x and then consider the behaviour of $u_0(t, x; k)$ as a function in t. By (3.55), for fixed x, we seek a t such that $u_0(t, x; k) \neq 0$, but the only candidate for t must have $R(t)^{-1}(x - X(t, y_0)) \in U$. For each point x that is close to the characteristic curve $\{X(s, y_0); s \in I\}$, there exists a point t_0 such that $u_0 \neq 0$ only in the interval of t centered at t_0 and which is determined by the size of $\operatorname{supp} u_0(0, \cdot)$. So, in this case u_0 becomes a function that oscillates with frequency $k/2\pi$. The above is a very general treatment; we shall discuss this again and in greater detail for the wave equation.

(c) The propagation of sound in air in which the temperature is not constant.

In Chapter 1, we obtained the equations (1.19) for a sound wave that propagates through air. We would now like to consider briefly the case for which at point x, in the domain where the sound propagates, the bulk modulus $K(x)$ is not fixed; i.e., the temperature is not constant. The case of constant $K(x)$ is dealt with in §3.5.

1. The surface temperature is cold while the air temperature is warm.

Let us take the earth's surface to correspond to the plane $x_3 = 0$ in $\mathbb{R}^3$, and the ambient air to be the region $x_3 > 0$. Further, if we assume that $K(x)$ is determined by its height above the earth's surface, then $K(x) = K(x_3)$. Since $K(x)$ is proportional to absolute temperature, we have that

$$(3.56) \qquad \frac{dK(x_3)}{dx_3} > 0.$$

With regard to the Hamiltonian $H(x, \tau, \xi) = \tau^2 - K(x_3)|\xi|^2$, the canonical equations are

$$\begin{cases} \dfrac{dT(s)}{ds} = 2\Theta(s), \quad \dfrac{d\Theta(s)}{ds} = 0, \\[2mm] \dfrac{dX_j(s)}{ds} = -2K(X_3(s))\Xi_j(s) \qquad (j = 1, 2, 3), \\[2mm] \dfrac{d\Xi_j(s)}{ds} = 0 \ (j = 1, 2), \quad \dfrac{d\Xi_3(s)}{ds} = \dfrac{\partial K}{\partial x_3}(X_3(s))|\Xi(s)|^2. \end{cases}$$

Without loss of generality, if we allow the origin of the sound to be at $x = 0$, then $X(0) = 0$. Also, let us suppose that $\Theta(0) = \sqrt{K(0)}|\Xi(0)|$ is satisfied. Since $K > 0$ and $T(s) = 2\Theta(0)s$, in the direction of sound conveyed in the air with time, we must have $\Xi_3(0) < 0$.

From (3.56) as s becomes large $\Xi_3(s)$ also increases, until finally it becomes positive. If $\Xi_3 > 0$, then over time $X_3(s)$ will decrease. Since

$\Xi_j(s) = \Xi_j(0)$ $(j = 1, 2)$, then there is no change in the horizontal direction.

Now, if we combine the behaviour of the solution of the canonical equations with Proposition 3.7, the way sound is transmitted is approximately as shown in the left-hand diagram of Figure 3.2. That is, when the earth's surface is cold while the air temperature is hot, even if there are buildings in its path, distant sound will often reach a far destination. In fact, from personal experience, it is possible to hear sound from the Osaka airport at the Mathematics Department of Osaka University, a distance of 3 km away. Usually, the sound of planes landing and taking off does not reach the department, however on occasions it feels as though the airport runway was right next door to the department. These occasions fitted in with the above described state.

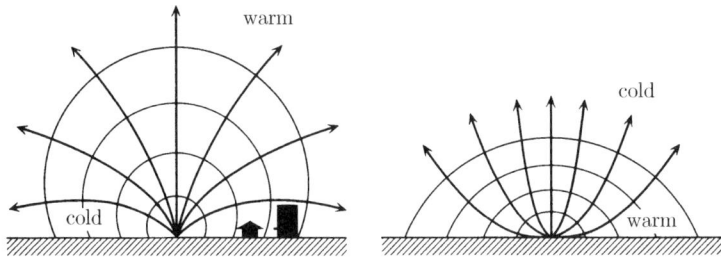

FIGURE 3.2

2. The surface temperature is warm while the air temperature is cold.

Instead of (3.56) we now have

$$(3.57) \qquad \frac{dK(x_3)}{dx_3} < 0.$$

Since this case is the exact opposite of that in the previous section, the propagation of sound is as in the right-hand diagram in Figure 3.2.

3.4 The propagation of singularities

In the previous section, we studied the propagation of the solution of the hyperbolic equation $Pu = 0$ with a high frequency. In this section, we shall again study the solution of $Pu = 0$, but here we shall consider the solution which has a singularity and then how this singularity propagates.

First, we assume that the singularity lies on a surface and so we ask, "What sort of characteristics must this surface possess?"

(a) The composition of a smooth function with a distribution.

Suppose that $\rho(t, x)$ is a smooth real-valued function defined on a domain V in $\mathbb{R}^{n+1}$. Further, suppose $S(s) \in \mathcal{D}'(\mathbb{R})$. Now, consider the distribution on V that substitutes ρ for s. For example, if $S(s) \in L^1_{\text{loc}}(\mathbb{R})$, then $S(\rho(t, x))$ is $L^1_{\text{loc}}(V)$. Hence, we can write $S(\rho)$ as an element of $\mathcal{D}'(V)$ given by this function, and so

$$(3.58) \qquad \langle S(\rho), \varphi \rangle_V = \int_V S(\rho(t, x)) \varphi(t, x) dt dx \qquad (\varphi \in \mathcal{D}(V)).$$

Next, assume that there is a domain $\widetilde{V}$ in $\mathbb{R}^{n+1}$, and

$$(3.59) \qquad V \ni (t, x) \longrightarrow (s, y) = (\rho(t, x), x) \in \widetilde{V}$$

is a diffeomorphism. Then we can assume the inverse mapping is given by

$$t = \mu(s, y), \quad x = y \qquad ((s, y) \in \widetilde{V}).$$

With the above in mind, we can express (3.58) as

$$(3.60) \qquad \begin{aligned} \int_V S(\rho(t, x)) \varphi(t, x) dt dx &= \int_{\widetilde{V}} S(s) \widetilde{\varphi}(s, y) J(s, y) ds dy \\ &= \int_{\mathbb{R}} S(s) \phi(s) ds. \end{aligned}$$

$$(3.61) \qquad \begin{aligned} \widetilde{\varphi}(s, y) &= \varphi(\mu(s, y), y), \qquad J(s, y) = \frac{\partial \mu}{\partial s}(s, y), \\ \phi(s) &= \int_{\mathbb{R}^n} \widetilde{\varphi}(s, y) J(s, y) dy. \end{aligned}$$

From (3.59), since $\varphi \in \mathcal{D}(V)$, we have that $\phi \in \mathcal{D}(\mathbb{R})$. Hence, we can express (3.58) as

$$\langle S(\rho), \varphi \rangle_V = \langle S(s), \phi(s) \rangle_{\mathbb{R}}.$$

With respect to the above, we give the following definition.

DEFINITION 3.8. Let us suppose that $S(s) \in \mathcal{D}'(\mathbb{R})$ and ρ is a function that is defined on a domain V in $\mathbb{R}^{n+1}$ and satisfies (3.59). Then, $S(\rho)$ is defined as an element of $\mathcal{D}'(V)$ that is determined by

$$(3.62) \qquad \langle S(\rho), \varphi \rangle_V = \langle S(s), \phi(s) \rangle_{\mathbb{R}} \qquad (\varphi \in \mathcal{D}(V)).$$

Here, $\phi \in \mathcal{D}(\mathbb{R})$ is determined from φ as follows:

$$\phi(s) = \int_{\mathbb{R}^{n-1}} \varphi(\mu(s,y), y) \frac{\partial \mu}{\partial s}(s,y) dy.$$

The next question to consider is, "With the above definition of $S(\rho)$ how can we express the partial differentials of this distribution?"

Now, from

$$t = \mu(\rho(t,x), x), \qquad \forall (t,x) \in V,$$

it follows that

$$(3.63) \qquad 1 = \frac{\partial \mu}{\partial s} \frac{\partial \rho}{\partial t}, \qquad 0 = \frac{\partial \mu}{\partial s} \frac{\partial \rho}{\partial x_j} + \frac{\partial \mu}{\partial y_j} \qquad (j = 1, 2, \dots, n).$$

On the other hand, since $\varphi(t,x) = \widetilde{\varphi}(\rho(t,x), x)$, we have

$$(3.64) \qquad \frac{\partial \varphi}{\partial x_j} J = \left(\frac{\partial \widetilde{\varphi}}{\partial s} \frac{\partial \rho}{\partial x_j} + \frac{\partial \widetilde{\varphi}}{\partial y_j} \right) J$$

$$(3.65) \qquad = \frac{\partial}{\partial s} \left(\widetilde{\varphi} \frac{\partial \rho}{\partial x_j} J \right) - \widetilde{\varphi} \frac{\partial}{\partial s} \left(\frac{\partial \rho}{\partial x_j} J \right) + \frac{\partial \widetilde{\varphi}}{\partial y_j} J.$$

By using the definition of J and then (3.63), we are led to the following relation:

$$\frac{\partial \rho}{\partial x_j} J = \frac{\partial \rho}{\partial x_j} \frac{\partial \mu}{\partial s} = -\frac{\partial \mu}{\partial y_j}.$$

From this relation we can obtain

$$-\widetilde{\varphi} \frac{\partial}{\partial s} \left(\frac{\partial \rho}{\partial x_j} J \right) + \frac{\partial \widetilde{\varphi}}{\partial y_j} J$$

$$= \widetilde{\varphi} \frac{\partial}{\partial s} \frac{\partial \mu}{\partial y_j} + \frac{\partial \widetilde{\varphi}}{\partial y_j} \frac{\partial \mu}{\partial s} = \frac{\partial}{\partial y_j} \left(\widetilde{\varphi} \frac{\partial \mu}{\partial s} \right).$$

Now, if we use this in the right-hand side of (3.64), we see that

$$\frac{\partial \varphi}{\partial x_j} J = \frac{\partial}{\partial s} \left(\widetilde{\varphi} \frac{\partial \rho}{\partial x_j} J \right) + \frac{\partial}{\partial y_j} \left(\widetilde{\varphi} \frac{\partial \mu}{\partial s} \right).$$

From the above, we obtain that

$$\int_U \frac{\partial \varphi}{\partial x_j}(\mu(s,y),y)J(s,y)dy$$

$$= \int_U \frac{\partial}{\partial s}\left\{\left(\widetilde{\varphi \frac{\partial \rho}{\partial x_j}}\right)(s,y)J(s,y)\right\}dy + \int_U \frac{\partial}{\partial y_j}\left(\widetilde{\varphi}\frac{\partial}{\partial s}\right)dy$$

$$= \frac{d}{ds}\int_U \left(\widetilde{\varphi \frac{\partial \rho}{\partial x_j}}\right)(s,y)J(s,y)dy.$$

Therefore, if we set

$$\phi_1(s) = \int_U \left(\widetilde{\varphi \frac{\partial \rho}{\partial x_j}}\right)(s,y)J(s,y),$$

then we have

$$\left\langle S(\rho), -\frac{\partial}{\partial x_j}\varphi\right\rangle_V = \left\langle S(s), -\frac{d}{ds}\phi_1(s)\right\rangle_{\mathbb{R}}$$

$$= \left\langle \frac{dS}{ds}(s), \phi_1(s)\right\rangle_{\mathbb{R}}.$$

From the definition of $\phi_1(s)$, we can write

$$\left\langle \frac{dS}{ds}(s), \phi_1(s)\right\rangle_{\mathbb{R}} = \left\langle \frac{dS}{ds}(\rho), \frac{\partial \rho}{\partial x_j}\varphi\right\rangle_V.$$

From the above we obtain

$$\left\langle \frac{\partial}{\partial x_j}S(\rho), \varphi\right\rangle_V = \left\langle \frac{dS}{ds}(\rho), \frac{\partial \rho}{\partial x_j}\varphi\right\rangle_V \qquad \forall \varphi \in \mathcal{D}(V).$$

Now, this equation shows that the following holds:

$$\frac{\partial}{\partial x_j}S(\rho) = \frac{dS}{ds}(\rho)\frac{\partial \rho}{\partial x_j}.$$

Also, using that $\dfrac{\partial \varphi}{\partial t}J = \dfrac{\partial \widetilde{\varphi}}{\partial s}$, we see that

$$\frac{\partial}{\partial t}S(\rho) = \frac{dS}{ds}(\rho)\frac{\partial \rho}{\partial t}.$$

Collecting the above together we obtain the following theorem.

THEOREM 3.9. *Suppose that $S \in \mathcal{D}'(\mathbb{R})$. Further, suppose that (3.59) gives a diffeomorphism from V to $I \times U$. If we define $S(\rho)$ as an element of $\mathcal{D}'(V)$ by (3.62), then for the partial differentials of $S(\rho)$ in $\mathcal{D}'(V)$ the following hold:*

$$(3.66) \qquad \frac{\partial}{\partial x_j} S(\rho) = \frac{dS}{ds}(\rho) \frac{\partial \rho}{\partial x_j}, \qquad \frac{\partial}{\partial t} S(\rho) = \frac{dS}{ds}(\rho) \frac{\partial \rho}{\partial t}.$$

Using (3.66), then for $v \in C^\infty(V)$ the Leibnitz formula gives
(3.67)

$$
\begin{aligned}
P(S(\rho)v) = & \frac{d^2 S}{ds^2}(\rho) \left((\rho_t)^2 + 2 \sum_{j=1}^n h_j \rho_{x_j} \rho_t - \sum_{j,l=1}^n a_{jl} \rho_{x_j} \rho_{x_l} \right) v \\
& + \frac{dS}{ds}(\rho) \left\{ 2 \left(\rho_t + 2 \sum_{j=1}^n h_j \rho_{x_j} \right) \frac{\partial v}{\partial t} \right. \\
& \qquad\qquad + 2 \sum_{j=1}^n \left(h_j \rho_t - \sum_{l=1}^n a_{jl} \rho_{x_l} \right) \frac{\partial v}{\partial x_j} \\
& \qquad\qquad\qquad \left. + (P - a_0)v \right\} + (Pv) S(\rho) \\
= & \frac{d^2 S}{ds^2}(\rho) H(t, x, \rho_t, \rho_x) v + \frac{dS}{ds}(\rho) K_\rho v + S(\rho)(Pv),
\end{aligned}
$$

where K_ρ denotes the differential operator obtained from (3.6) by replacing Φ by ρ.

(b) Propagation of discontinuities.

For the propagation phenomena governed by the equation $Pu = 0$ the question of interest is, "What sort of propagation has a singularity that is a discontinuity, say?" This problem includes the consideration of the phenomena obtained by giving a shock to some point, which is then transmitted to its surroundings.

In this section, as the easiest singularity to understand, we will deal with a discontinuity. In this case, i.e., a discontinuity that lies on a smooth surface, how is it transmitted with time?

Let us suppose that $\rho(t, x)$ is a smooth real-valued function defined on the open set V and which satisfies

$$(3.68) \qquad |\rho_t(t, x)|^2 + |\rho_x(t, x)|^2 \neq 0.$$

Further, suppose that $(t_0, x_0) \in V$ and that $\rho(t_0, x_0) = 0$. From the assumption in (3.68), $\mathcal{C} = \{(t, x); \rho(t, x) = 0\}$ is a smooth surface that contains the point (t_0, x_0). If $|\rho_x| \neq 0$, for each point of t, $\mathcal{C}(t) = \{x; \rho(t, x) = 0\}$ is a smooth surface in $\mathbb{R}^n$.

Let us now consider the solution of $Pu = 0$ that has the discontinuous surface $\mathcal{C} = \{(t, x); \rho(t, x) = 0\}$ in the (t, x)-space. For example, suppose for some positive constant α, the solution u satisfies

$$(3.69) \qquad \begin{cases} |u| \geqslant \alpha & \text{in } \{(t, x); \rho(t, x) > 0\} = V^+, \\ u = 0 & \text{in } \{(t, x); \rho(t, x) < 0\} = V^-, \end{cases}$$

and u is C^∞ on $\overline{V^+}$. When such a u satisfies $Pu = 0$ as a distribution, we would like to know what sort of properties the discontinuous surface $\mathcal{C}$ possesses?

Since $u \in C^\infty(\overline{V^+})$, there is a $v \in C^\infty(V)$ such that

$$(3.70) \qquad\qquad u = Y(\rho)v,$$

where Y is the Heaviside function. Now, by applying this to (3.67) we obtain

$$(3.71) \qquad Pu = \frac{d^2 Y}{ds^2}(\rho) H(t, x, \rho_x, \rho_y) + \frac{dY}{ds}(\rho) K_\rho v + Y(\rho)(Pv).$$

Now, take $\psi(s) \in \mathcal{D}(\mathbb{R})$ and $\eta(y) \in \mathcal{D}(U)$; then for $0 < \varepsilon \ll 1$ let

$$\varphi_\varepsilon(t, x) = \psi\left(\frac{\rho}{\varepsilon}\right)\eta(y).$$

Further, let $w_0 = H(t, x, \rho_t, \rho_x)v$, $w_1 = K_\rho v$ and $w_2 = Pv$, and so we set

$$(3.72) \qquad \phi_j(s) = \int_U \widetilde{w}_j(s, y)\eta(y)J(s, y)\,dy \qquad (j = 0, 1, 2).$$

Then, we can see that

$$\left\langle \frac{d^2 Y}{ds^2}(\rho)w_0, \varphi_\varepsilon \right\rangle_V = \left\langle \frac{d^2 Y}{ds^2}(s), \psi\left(\frac{s}{\varepsilon}\right)\phi_0(s) \right\rangle_{\mathbb{R}} \quad (= I_\varepsilon).$$

Now, since $\dfrac{d^2 Y}{ds^2}(s) = \delta'(s)$, we obtain that

$$I_\varepsilon = -\frac{d}{ds}\left(\psi\left(\frac{s}{\varepsilon}\right)\phi_0(s)\right)\Big|_{s=0} = -\frac{1}{\varepsilon}\psi'(0)\phi_0(0) - \psi(0)\frac{d\phi_0}{ds}(0).$$

By similar means, we may write

$$\left\langle \frac{dY}{ds}(\rho)K_\rho v, \varphi_\varepsilon \right\rangle_V = \left\langle \frac{dY}{ds}(s), \psi\left(\frac{s}{\varepsilon}\right)\phi_1(s) \right\rangle_{\mathbb{R}} \quad (= \mathrm{II}_\varepsilon).$$

Therefore,

$$\mathrm{II}_\varepsilon = \left\langle \frac{dS}{ds}(\rho)v_1, \varphi_\varepsilon \right\rangle = \psi(0)\phi_1(0).$$

Also,

$$\langle Y(s)Pv, \varphi_\varepsilon \rangle_V = \left\langle Y(s), \psi\left(\frac{s}{\varepsilon}\right)\phi_2(s) \right\rangle_{\mathbb{R}} \quad (= \mathrm{III}_\varepsilon).$$

But, since $\phi_2 \in \mathcal{D}(\mathbb{R})$,

(3.73) $|\mathrm{III}_\varepsilon| \leqslant C\varepsilon.$

Now choose ψ so that $\psi'(0) = 1$, and suppose that $w_0(\widetilde{t}, \widetilde{x}) \neq 0$ at $(\widetilde{t}, \widetilde{x}) \in \mathcal{C}$. From the fact that $J(s, y) \neq 0$, by suitably choosing $\eta \in \mathcal{D}(U)$, we can deduce that $\phi_0(0) \neq 0$. Then, for such ψ and η, we obtain that

(3.74) $\langle Pu, \varphi_\varepsilon \rangle = \mathrm{I}_\varepsilon + \mathrm{II}_\varepsilon + \mathrm{III}_\varepsilon,$

but since

$$|\mathrm{I}_\varepsilon| \geqslant \frac{1}{\varepsilon}|\phi_0(0)| - C, \quad |\mathrm{II}_\varepsilon| \leqslant C, \quad |\mathrm{III}_\varepsilon| \leqslant C_\varepsilon,$$

we see that $Pu = 0$ never holds in $\mathcal{D}'(V)$. Therefore, for the u given in (3.70) to satisfy $Pu = 0$ we must have that for all $(t, x) \in \mathcal{C}$

$$w_0(t, x) = H(t, x, \rho_t, \rho_x)v(t, x) = 0.$$

Before we state the next proposition, we note that from (3.69) it follows that $v(t, x) \neq 0$ on $\mathcal{C}$.

PROPOSITION 3.10. *Suppose the following: u satisfies (3.69), u is a C^∞-function on $\overline{V^+}$, and $Pu = 0$ as a distribution. Then the function ρ that defines a discontinuous surface $\mathcal{C}$ of u must satisfy*

(3.75) $H(t, x, \rho_t, \rho_x) = 0, \qquad \forall (t, x) \in \mathcal{C}.$

Now, we assume that ρ satisfies (3.75). Then, from (3.74) we see that

$$\langle Pu, \varphi_\varepsilon \rangle = \psi(0)\left(-\frac{d\phi_0}{ds}(0) + \phi_1(0)\right) + \mathrm{III}_\varepsilon.$$

If we choose ψ so that $\psi(0) \neq 0$ and consider (3.73), then for $Pu = 0$ to hold we must have that

(3.76) $$-\frac{d\phi_0}{ds}(0) + \phi_1(0) = 0.$$

Consider (3.72) with $j = 0$. Then, since $H|_{s=0} = 0$, we can write

$$\frac{d\phi_0}{ds}(0) = \int_U \left[\frac{d}{ds} H(t, x, \rho_t, \rho_x)\right]_{s=0} \widetilde{v}(0, y)\psi(y)J(0, y)dy.$$

Therefore, if we set

$$\widetilde{\kappa}(y) = -\frac{d}{ds} H(t, x, \rho_t, \rho_x)\bigg|_{s=0},$$

then (3.76) becomes

$$\int_U \{(K_\rho v)\widetilde{\ }(0, y) + \widetilde{\kappa}(y)\widetilde{v}(0, y)\}\psi(y)J(0, y)dy = 0.$$

Since the above formula holds for all $\psi \in \mathcal{D}(U)$, we see that

(3.77) $$(K_\rho v)(t, x) + \kappa(t, x)v(t, x) = 0 \qquad \text{on } \mathcal{C},$$

where κ is a function defined on $\mathcal{C}$ by $\kappa(\mu(0, y), y) = \widetilde{\kappa}(y)$. If we have $H(t, x, \rho_t, \rho_x) = 0$ in V, then $\kappa \equiv 0$.

Now, we confirm (3.77) is an equation on $\mathcal{C}$. To this end, let

$$k_0 = 2\left(\frac{\partial\rho}{\partial t} + \sum_{l=1}^n h_l \frac{\partial\rho}{\partial x_l}\right), \quad k_j = 2\left(h_j \frac{\partial\rho}{\partial t} - 2\sum_{l=1}^n a_{jl} \frac{\partial\rho}{\partial x_l}\right),$$

$$m = (P - a_0)\rho.$$

Then, we can write

$$K_\rho v = k_0 \frac{\partial v}{\partial t} + \sum_{j=1}^n k_j \frac{\partial v}{\partial x_j} + mv.$$

So, let

$$\widetilde{v}(s, y) = v(\mu(s, y), y);$$

then

$$\frac{\partial\widetilde{v}}{\partial y_j} = \frac{\partial v}{\partial t} \frac{\partial\mu}{\partial y_j} + \frac{\partial v}{\partial x_j}.$$

Therefore,

$$\sum_{j=1}^{n} k_j \frac{\partial \tilde{v}}{\partial y_j} = \sum_{j=1}^{n} k_j \frac{\partial \mu}{\partial y_j} \frac{\partial v}{\partial t} + \sum_{j=1}^{n} k_j \frac{\partial v}{\partial x_j}.$$

But, from (3.63), we know that $\dfrac{\partial \mu}{\partial y_j} = -\left(\dfrac{\partial \rho}{\partial t}\right)^{-1} \dfrac{\partial \rho}{\partial x_j}$, and so

$$\sum_{j=1}^{n} k_j \frac{\partial \mu}{\partial y_j} = 2\left(\frac{\partial \rho}{\partial t}\right)^{-1} \left\{ -\sum_{j=1}^{n} h_j \frac{\partial \rho}{\partial t} \frac{\partial \rho}{\partial x_j} + \sum_{j,l=1}^{n} a_{jl} \frac{\partial \rho}{\partial x_j} \frac{\partial \rho}{\partial x_l} \right\}$$

$$= 2\left(\frac{\partial \rho}{\partial t}\right)^{-1} \left\{ -H(t,x,\rho_t,\rho_x) + \left(\frac{\partial \rho}{\partial t}\right)^2 + \sum_{j=1}^{n} h_j \frac{\partial \rho}{\partial t} \frac{\partial \rho}{\partial x_j} \right\}$$

$$= 2\left(\frac{\partial \rho}{\partial t} + \sum_{j=1}^{n} h_j \frac{\partial \rho}{\partial x_j}\right) = k_0.$$

That is, we can express

$$\sum_{j=1}^{n} k_j \frac{\partial \tilde{v}}{\partial y_j} = k_0 \frac{\partial v}{\partial t} + \sum_{j=1}^{n} k_j \frac{\partial v}{\partial x_j} = K_\rho v - mv.$$

From the above we now have that (3.77) is equivalent to the following:

$$\sum_{j=1}^{n} \tilde{k}_j(0,y) \frac{\partial \tilde{v}}{\partial y_j} + \tilde{m}(0,y)\tilde{v}(0,y) + \tilde{\kappa}(y)\tilde{v}(0,y) = 0;$$

i.e., (3.77) is a differential equation on $\mathcal{C}$.

Before stating the next proposition, we note that $\tilde{v}(0,y)$ is the size of the gap on the discontinuous surface $\mathcal{C}$ of u given by (3.69).

PROPOSITION 3.11. *Suppose that the same assumptions hold for u as in Proposition 3.10. Also, suppose that the defining function ρ on the discontinuous surface $\mathcal{C}$ satisfies (3.75). Further, we denote the gap on the discontinuous surface $\mathcal{C}$ of u by $[u]$. Then, $[u]$ must satisfy equation (3.77) on the surface $\mathcal{C}$.*

Further, if the defining function ρ satisfies the following condition stronger than (3.77), namely,

(3.78) $$H(t,x,\rho_t,\rho_x) = 0 \quad \forall (t,x) \in V,$$

then [*u*] *must satisfy*

$$K_\rho[u] = 0 \quad on \ C.$$

(c) The propagation of a general singularity.

Although the reader may have already noticed, it is pertinent to reiterate that during the investigation of the solutions with discontinuity, we use the *Eikonal* equation for the principal part of P, the transport equation and, further, the same discussion as for the construction of the asymptotic solution studied in §3.1. So, using the same methods as for the construction of the asymptotic solution, we develop a construction for a solution that has a discontinuity or a stronger singularity on the surface.

To begin with, suppose that $S_j \in \mathcal{D}'(\mathbb{R})$ ($j = 0, 1, 2, \ldots$) satisfies

$$(3.79) \qquad \qquad \operatorname{supp} S_j \subset [0, \infty)$$

and

$$(3.80) \qquad \qquad \frac{dS_{j+1}}{ds} = S_j \qquad (j = 0, 1, 2, \ldots).$$

Further, we take $\rho(t, x) \in C^\infty(V)$ to satisfy

$$(3.81) \qquad \qquad H(t, x, \rho_t, \rho_x) = 0, \quad \forall (t, x) \in V.$$

Then we seek a $u(t, x)$ of the form

$$(3.82) \qquad \qquad u(t, x) = \sum_{j=0}^{N} S_j(\rho) v_j(t, x)$$

such that $Pu = 0$.

As noted previously, if u is a solution with a discontinuity, it is sufficient to set

$$(3.83) \qquad \qquad S_j(s) = \frac{1}{j!} s^j Y(s) \quad (j = 0, 1, 2, \ldots).$$

But, if u has a higher singularity, then, for an integer $m \geqslant 0$, we take

$$(3.84) \qquad \begin{cases} S_j(s) = \delta^{(m-j)}(s) \quad (j = 0, 1, 2, \ldots, m), \\[2mm] S_j(s) = \dfrac{s^{j-(m+1)}}{(j-m-1)!} Y(s) \quad (j = m+1, m+2, \ldots). \end{cases}$$

As a further example, we take $F(s) \in C_0^\infty(0, \infty)$ and for $\varepsilon > 0$ we set

$$S_0^{(\varepsilon)}(s) = \frac{1}{\varepsilon} F\left(\frac{s}{\varepsilon}\right),$$

$$S_j^{(\varepsilon)}(s) = \int_0^s \frac{(s-\sigma)^{j-1}}{(j-1)!} \frac{1}{\varepsilon} F\left(\frac{\sigma}{\varepsilon}\right) d\sigma \quad (j = 1, 2, \dots).$$

If $\displaystyle\int_{\mathbb{R}} F(s)ds = 1$, then for S_j of (3.84) with $m = 0$, the following holds:

$$S_j^{(\varepsilon)}(s) \longrightarrow S_j(s) \quad \text{in } \mathcal{D}'(\mathbb{R}).$$

But, this is just an approximation of (3.84).

Now, look at the operation P for u of the form in (3.82). By applying (3.67) to each $S_j(\rho)v_j$, and noting the hypothesis in (3.81), then from (3.80) we have

$$P(S_0(\rho)v_0) = \frac{dS_0}{ds}(\rho)K_\rho v_0 + S_0(\rho)Pv_0,$$

$$P(S_j(\rho)v_j) = S_{j-1}(\rho)K_\rho v_j + S_j(\rho)Pv_j \quad (j = 1, 2, \dots, N).$$

Therefore, if v_j $(j = 1, 2, \dots)$ satisfies

(3.85) $$\begin{cases} K_\rho v_0 = 0 & \text{in } V, \\ K_\rho v_j = -Pv_{j-1} & \text{in } V \quad (j = 1, 2, \dots, N), \end{cases}$$

then

(3.86) $$Pu = S_N(\rho)v_N.$$

From the above examples and also from (3.80) it is easy to see that as j becomes large in accompaniment S_j gradually becomes smooth. Therefore, if we choose N to be sufficiently large, then $S_N(s)$ is reasonably smooth, and so $S_N(\rho(t, x))$ can also be thought of as a smooth function in (t, x). Hence, the right-hand side of (3.86) becomes sufficiently smooth.

(d) The solution for a discontinuous initial value.

We begin by considering the following initial value problem:

(3.87) $$\begin{cases} Pu = 0, \\ u(0, x) = 0, \quad \dfrac{\partial u}{\partial t}(0, x) = u_1(x), \end{cases}$$

where we have assumed that $u_1(x)$ is a function with a discontinuity as given below. But, first we assume that $\rho_0(x)$ is a real-valued smooth function defined in some neighbourhood, U, of x_0 and satisfies

(3.88)
$$\begin{cases} \sum_{j=1}^{n} |(\rho_0)_{x_j}(x)|^2 \neq 0, \quad \forall x \in U, \\ \rho_0(x_0) = 0. \end{cases}$$

If we set $\mathcal{C}_0 = \{x \in U; \rho_0(x) = 0\}$, then $\mathcal{C}_0$ is a smooth surface that passes through x_0. Now, let u_1 be a function defined by

(3.89)
$$u_1(x) = Y(\rho_0(x))v(x),$$

where v is a $C^\infty(U)$-function. So, u_1 is a function that has a discontinuity on $\mathcal{C}_0$.

From (3.88), if we define two real-valued functions $\theta^\pm$ on U by $\theta^\pm(x) = \lambda^\pm(0, x, (\rho_0)_x)$, then the following holds:

(3.90)
$$H(0, x, \theta^\pm(x), (\rho_0)_x(x)) = 0 \quad \forall x \in U.$$

From the results in §3.2, for some neighbourhood V of $(0, x_0)$, there exists a smooth function $\rho^\pm(t, x)$ defined on V that satisfies

$$\begin{cases} H(t, x, \rho_t^\pm, \rho_x^\pm) = 0, \quad \forall (t, x) \in V, \\ \rho^\pm(0, x) = \rho_0(x), \quad \rho_t^\pm(0, x) = \theta^\pm(x). \end{cases}$$

Given S_j by (3.83), we shall seek an approximate solution of (3.87) of the form

(3.91)
$$u_N = \sum_{j=1}^{N} S_j(\rho^+)v_j^+ + \sum_{j=1}^{N} S_j(\rho^-)v_j^- = u^+ + u^-.$$

We begin by considering the initial conditions, in fact, they are

$$u(0, x) = \sum_{j=1}^{N} S_j(\rho_0(x))v_j^+ + \sum_{j=1}^{N} S_j(\rho_0(x))v_j^-,$$

$$\frac{\partial u}{\partial t}(0, x) = \sum_{j=1}^{N} \left(\frac{dS_j}{ds}(\rho_0(x))\theta^+ v_j^+ + S_j(\rho_0(x))\frac{\partial v_j^+}{\partial t} \right)$$

$$+ \sum_{j=1}^{N} \left(\frac{dS_j}{ds}(\rho_0(x))\theta^- v_j^- + S_j(\rho_0(x))\frac{\partial v_j^-}{\partial t} \right).$$

If we note that $\dfrac{dS_1}{ds} = Y$, in order to satisfy the initial conditions, at $t = 0$ we desire that the following hold:

$$(3.92) \quad \begin{cases} v_j^+ + v_j^- = 0 \quad (j = 1, 2, \ldots, N), \\ \theta^+ v_1^+ + \theta^- v_1^- = v(x), \\ \theta^+ v_j^+ + \theta^- v_j^- + \dfrac{\partial v_{j-1}^+}{\partial t} + \dfrac{\partial v_{j-1}^-}{\partial t} = 0 \quad (j = 2, 3, \ldots, N). \end{cases}$$

First, choose $v_0^\pm|_{t=0}$ such that

$$(3.93) \qquad v_1^+(0, x) = \frac{1}{\theta^+ - \theta^-} v(x), \quad v_1^-(0, x) = -\frac{1}{\theta^+ - \theta^-} v(x).$$

Then assume $v_1^\pm$ is a solution of

$$K_{\rho^+} v_1^+ = 0, \quad K_{\rho^-} v_1^- = 0 \quad \text{in } V,$$

that satisfies (3.93). From the above we can determine $v_1^\pm$.

Next, let the value of $v_2^\pm$ at $t = 0$ be

$$v_2^\pm(0, x) = \mp \frac{1}{\theta^+ - \theta^-} \left(\frac{\partial v_1^+}{\partial t} + \frac{\partial v_1^-}{\partial t} \right)(0, x),$$

and let $v_2^\pm$ be the solution of

$$K_{\rho^\pm} v_2^\pm = P v_1^\pm.$$

Repeating the above procedure, we obtain $v_j^\pm$ $(j = 1, 2, \ldots, N)$.

For the $v_j^\pm$ determined as above, u given by (3.91) satisfies the following:

$$\begin{cases} Pu = S_N(\rho^+) P v_N^+ + S_N(\rho^-) P v_N^-, \\ u(0, x) = 0, \quad \dfrac{\partial u(0, x)}{\partial t} = u_1(x) + S_N(\rho_0) \left(\dfrac{\partial v_N^+}{\partial t} + \dfrac{\partial v_N^-}{\partial t} \right). \end{cases}$$

Now, suppose that $w(t, x)$ is a solution of

$$\begin{cases} Pw = -Pu, \\ w(0, x) = 0, \quad \dfrac{\partial w}{\partial t}(0, x) = -S_N(\rho_0) \left(\dfrac{\partial v_N^+}{\partial t} + \dfrac{\partial v_N^-}{\partial t} \right). \end{cases}$$

Since we have that

$$Pu \in C^{N-1}(V) \quad \text{and} \quad \frac{\partial w}{\partial t}(0, x) \in C^{N-1}(U),$$

we see that $w \in C^{N-[(n+1)/2]}(V)$. Clearly, we have that

$$P(u + w) = 0.$$

A strict solution of (3.87) is established from $u + w$, but the discontinuity comes from u. Therefore, we have shown that the discontinuities of the strict solution lie on the two characteristic surfaces $\rho^{\pm}(t, x) = 0$ that contain $\mathcal{C}_0$.

3.5 The asymptotic solution of the wave equation and geometrical optics

In this section, we would like to consider in more detail the properties of the asymptotic solution for the wave equation of 3-dimensional space. In this respect, we would like to consider the propagation of the solution in a region with a boundary and the propagation of the solution for the case when two types of media come into contact. The behaviour of the solution in these problems explains the reflection on the boundary of light and sound and the phenomenon of refraction of light that occurs when it enters water from air.

If we consider the high frequency case and construct asymptotic solutions corresponding to these problems, then this yields properties of light that are known as *geometrical optics*.

(a) The properties of the solution of the Eikonal equation.

In this subsection, we consider the asymptotic solution of the wave equation. That is, we consider the asymptotic solution of $Pu = 0$ with

$$P = \frac{\partial^2}{\partial t^2} - v^2 \Delta, \quad \Delta = \sum_{j=1}^{3} \frac{\partial^2}{\partial x_j^2}.$$

As explained in §1.3, since we can reduce our deliberations to the case of $v = 1$ by a change of variable of t, in what follows we will concern ourselves solely with the case $v = 1$. Therefore, we denote, hereafter, the operator by $\square$ rather than by P. Also, for this case the Hamiltonian is $H = \tau^2 - |\xi|^2$.

Now, suppose that φ is a smooth real-valued function defined on an open set Ω in $\mathbb{R}^3$ and satisfies

$$(3.94) \qquad |\nabla\varphi(x)|^2 = \sum_{j=1}^{3} \left(\frac{\partial\varphi}{\partial x_j}(x) \right)^2 = 1 \quad (x \in \Omega).$$

Next, if we set

(3.95) $$\Phi(t,x) = t - \varphi(x),$$

then Φ is such that $H(\Phi_t, \Phi_x) = 0$. So, the set given by "$\Phi(t,x) =$ a constant" is a smooth surface in the (t,x)-space and, further, is the characteristic surface of P. So, in this section, we restrict our attention to the asymptotic solution of the form

(3.96) $$u(t,x;k) = e^{ik(t-\varphi(x))}v(t,x;k),$$

with

(3.97) $$v(t,x;k) = \sum_{j=0}^{m+N} (ik)^{m-j}v_j(t,x).$$

Clearly, since $\square$ is one of the examples of the operators that were studied in §3.1 through §3.3, the asymptotic solution has already been constructed. However, for the case of $\square$ it is quite straightforward to see the properties of the asymptotic solution in more detail.

Now, the canonical equations for the Hamiltonian $H(\tau,\xi) = \tau^2 - |\xi|^2$ are

(3.98) $$\begin{cases} \dfrac{\partial T(s)}{\partial s} = 2\Theta(s), & \dfrac{\partial \Theta(s)}{\partial s} = 0, \\ \dfrac{\partial X(s)}{\partial s} = -2\Xi(s), & \dfrac{\partial \Xi(s)}{\partial s} = 0. \end{cases}$$

Therefore, $\Theta(s) = \Theta(0)$ and $\Xi(s) = \Xi(0)$. Suppose that Φ is given by (3.95), since Φ satisfies the *Eikonal equation*, if we set $T(0) = t_0$, $X(0) = y$, $\Theta(0) = \Phi_t = 1$ and $\Xi(0) = \Phi_x(t_0,y) = -\nabla(t_0,y) = -\nabla\varphi(y)$, then

$$X(s) = y + 2s\nabla\varphi(y), \quad T(s) = t_0 + 2s.$$

In addition, from (3.41) of Theorem 3.5 we have that $\Xi(s) = -(\nabla\varphi)(X(s))$, and so we see that

(3.99) $$(\nabla\varphi)(y + 2s\nabla\varphi(y)) = (\nabla\varphi)(y).$$

(b) The transport equation of the wave equation.

Suppose that $\varphi(x)$ is a smooth real-valued function, defined on $\Omega \subset \mathbb{R}^3$, that satisfies (3.94). Hereafter, we would like to consider our problem in the interior of the Ω. Now, let

$$(3.100) \qquad \mathcal{C}_\varphi(x) = \{y; \varphi(y) = \varphi(x)\}.$$

From (3.94), $\mathcal{C}_\varphi(x)$ is a smooth surface that passes through x. This surface we will call the *wave front* of the function given by (3.96). In the case when k is large in comparison to the rate of change of v, then the graph of $\operatorname{Re} u(t, x; k)$ for the function in x with fixed t is as in Figure 3.3.

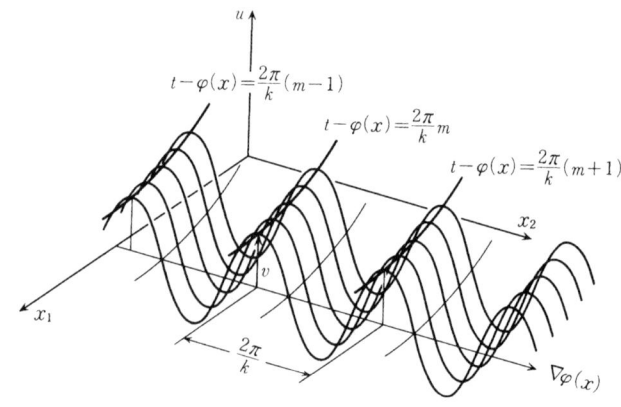

FIGURE 3.3

The position of the troughs and peaks of the wave can be expressed as a surface of the form $\{y; \varphi(y) = a\ constant\,\}$. So, with time t, we will see shortly that the position of each of the peaks moves with speed 1 in the direction $\nabla\varphi$. Since what we can recognize as a usual wave front is the surface that forms the peaks of the wave or the troughs of the wave, it is natural to call $\mathcal{C}_\varphi(x)$ the wave front that passes through x of (3.96).

It may seem obvious, but we want to confirm that $\nabla\varphi$ is the unit normal vector of $\mathcal{C}_\varphi(x)$.

In fact, since $\mathcal{C}_\varphi(x)$ is a smooth surface, and if we consider a neighbourhood U of x in $\mathbb{R}^3$, then we can find a smooth function, defined on a neighbourhood V of the origin of $\mathbb{R}^2$,

$$V \ni \sigma = (\sigma_1, \sigma_2) \longrightarrow s(\sigma) = (s_1(\sigma), s_2(\sigma), s_3(\sigma)) \in \mathbb{R}^3$$

that takes its value on $\mathbb{R}^3$. Further, $s(0) = x$ and

$$\frac{\partial s}{\partial \sigma_1}(\sigma) \text{ and } \frac{\partial s}{\partial \sigma_2}(\sigma) \text{ are linearly independent.}$$

So, we can write

(3.101) $$\mathcal{C}_\varphi(x) \cap U = \{s(\sigma); \sigma \in V\}.$$

The tangent plane at $s(\sigma)$ of $\mathcal{C}_\varphi(x)$ is given by

$$T_s\mathcal{C}_\varphi(x) = \left\{ a_1 \frac{\partial s}{\partial \sigma_1}(\sigma) + a_2 \frac{\partial s}{\partial \sigma_2}(\sigma); a_1, a_2 \in \mathbb{R} \right\}.$$

Since by the definition of $\mathcal{C}_\varphi(x)$, $\varphi(s(\sigma)) = \varphi(x)$ $(\sigma \in V)$, then by partially differentiating both sides with respect to σ_1 and σ_2, we obtain

(3.102) $$\sum_{j=1}^{3} \frac{\partial \varphi}{\partial x_j}(s(\sigma)) \frac{\partial s_j}{\partial \sigma_l}(\sigma) = 0 \quad (l = 1, 2).$$

Since the left-hand side of (3.102) can be expressed as $\nabla\varphi(s) \cdot \frac{\partial s}{\partial \sigma_l}(\sigma)$, $\nabla\varphi(s)$ is orthogonal to all the elements of $T_s\mathcal{C}_\varphi(x)$. That is, $\nabla\varphi(s)$ is the unit normal vector of $\mathcal{C}_\varphi(x)$ at $s \in \mathcal{C}_\varphi(x)$.

The operator K for the transport equation is

(3.103) $$K = 2\frac{\partial}{\partial t} + 2\sum_{j=1}^{3} \frac{\partial \varphi}{\partial x_j}(x) \frac{\partial}{\partial x_j} + \Delta\varphi(x).$$

Hence, in order to study the solution of the transport equation, we need to know the value of $\Delta\varphi$, and this is taken care of by the following theorem.

THEOREM 3.12. *For φ that satisfies (3.94) we have that*
(3.104)
$$\Delta\varphi(x) = 2 \times \text{ the average curvature of the surface } \mathcal{C}_\varphi(x) \text{ at } x,$$

where we suppose that the average curvature is taken in the direction $\nabla\varphi(x)$.

The proof of the above theorem is dealt with in sections (c), (d) and (e) below. In fact, sections (c) and (d) deal with fundamental concepts of geometry. The discussions and explanation, therein, are geared towards those who might not be so familiar with these concepts.

(c) Curvature.

In this section, we give a definition of the curvature of a surface. So, let us begin by setting $\nu(\sigma) = \nabla\varphi(s(\sigma))$. Since $\nu(\sigma) \cdot \nu(\sigma) = 1$ from (3.94), by partially differentiating both sides of (3.94) with respect to σ_l $(l = 1, 2)$, we obtain

$$2\nu(\sigma) \cdot \frac{\partial \nu}{\partial \sigma_l}(\sigma) = 0.$$

The above formula gives that $\dfrac{\partial \nu}{\partial \sigma_l}(s)$ is orthogonal to the normal vector $\nu(\sigma)$; i.e., the formula shows that $\dfrac{\partial \nu}{\partial \sigma_l}(s)$ belongs to $T_s\mathcal{C}_\varphi(x)$. Therefore, we can write

$$(3.105) \qquad \frac{\partial \nu}{\partial \sigma_l}(\sigma) = \sum_{j=1}^{2} \kappa_{lj}(\sigma) \frac{\partial s}{\partial \sigma_j}(\sigma) \quad (l = 1, 2).$$

Now, we consider the eigenvalues of the following 2×2 real matrix:

$$K(\sigma) = \begin{bmatrix} \kappa_{11}(\sigma) & \kappa_{12}(\sigma) \\ \kappa_{21}(\sigma) & \kappa_{22}(\sigma) \end{bmatrix}.$$

First, we note that if we let $\tau_l(\sigma) = \dfrac{\partial s}{\partial \sigma_l}(\sigma)$, then we can write

$$\nu(\sigma) = \pm \frac{\tau_1(\sigma) \times \tau_2(\sigma)}{|\tau_1(\sigma) \times \tau_2(\sigma)|}.$$

So, we look at the positive case. Then, by a direct calculation, we see that

$$\frac{\partial \nu}{\partial \sigma_1} \cdot \tau_2 = \frac{1}{|\tau_1 \times \tau_2|} \left(\tau_1 \times \frac{\partial \tau_2}{\partial \sigma_1} \right) \cdot \tau_2,$$

$$\frac{\partial \nu}{\partial \sigma_2} \cdot \tau_1 = \frac{1}{|\tau_1 \times \tau_2|} \left(\frac{\partial \tau_1}{\partial \sigma_2} \times \tau_2 \right) \cdot \tau_1.$$

But since $\dfrac{\partial \tau_2}{\partial \sigma_1} = \dfrac{\partial^2 s}{\partial \sigma_1 \partial \sigma_2} = \dfrac{\partial \tau_1}{\partial \sigma_2}$, the above two quantities are equal.

On the other hand, using (3.105), we have

$$(3.106) \qquad \left(\frac{\partial \nu}{\partial \sigma_l} \cdot \tau_j \right)_{l,j=1,2} = KT, \qquad T = [\tau_j \cdot \tau_k]_{j,k=1,2}.$$

From our previous note, it is easy to see that the matrix in the left-hand side of the above is a symmetric matrix. On the other hand, T is not only a symmetric matrix but also a positive matrix. Therefore, $T^{1/2}$ can be defined, and hence from (3.106) we get

$$(3.107) \quad T^{-1/2}KT^{1/2} = T^{-1/2}NT^{-1/2}, \quad N = \left(\frac{\partial \nu}{\partial \sigma_l} \cdot \tau_j \right)_{l,j=1,2}.$$

Since the right-hand side of the above is a symmetric matrix, the eigenvalues are real. Also, from its form, the eigenvalues of the matrix of the left-hand side agree with those of K. Therefore, the eigenvalues of K are the same as those of the right-hand side of (3.107) and both are real.

Now, we will show that the eigenvalues of K do not depend on the way we express $\mathcal{C}_\varphi(x)$, that is, these eigenvalues are numbers with certain geometric properties. With this purpose in mind, we change the way we express $\mathcal{C}_\varphi(x)$.

First, we assume that $\widetilde{V}$ is a neighbourhood of the origin in $\mathbb{R}^2$, and then suppose that g is a diffeomorphism from $\widetilde{V}$ to V. Further, we use the change of variable $\sigma = g(\widetilde{\sigma})$, where $\widetilde{\sigma} = (\widetilde{\sigma}_1, \widetilde{\sigma}_2) \in \widetilde{V}$. Next, let $\widetilde{s}(\widetilde{\sigma}) = s(g(\widetilde{\sigma}))$ and $\widetilde{\nu}(\widetilde{\sigma}) = \nu(g(\widetilde{\sigma}))$. From the formula of the partial differentials of the composite function, we obtain that
(3.108)

$$\frac{\partial \widetilde{\nu}}{\partial \widetilde{\sigma}_l}(\widetilde{\sigma}) = \sum_{k=1}^{2} \frac{\partial \nu}{\partial \sigma_k}(\sigma)\frac{\partial \sigma_k}{\partial \widetilde{\sigma}_l}(\widetilde{\sigma}), \qquad \frac{\partial \widetilde{s}}{\partial \widetilde{\sigma}_k}(\widetilde{\sigma}) = \sum_{j=1}^{2} \frac{\partial s}{\partial \sigma_j}(\sigma)\frac{\partial \sigma_j}{\partial \widetilde{\sigma}_k}(\widetilde{\sigma}).$$

Now set

$$(3.109) \qquad \frac{\partial \widetilde{\nu}}{\partial \widetilde{\sigma}_l}(\widetilde{\sigma}) = \sum_{j=1}^{2} \widetilde{\kappa}_{lk}(\widetilde{\sigma})\frac{\partial \widetilde{s}}{\partial \widetilde{\sigma}_k}(\widetilde{\sigma}).$$

By substituting the first formula of (3.108) into the left-hand side of (3.109) and then using (3.105), while for the right-hand side of (3.109) by applying the second formula of (3.108), we see that

$$\sum_{k=1}^{2} \left(\sum_{j=1}^{2} \kappa_{kj}(\sigma)\frac{\partial s}{\partial \sigma_j}(\sigma) \right) \frac{\partial \sigma_k}{\partial \widetilde{\sigma}_l} = \sum_{k=1}^{2} \widetilde{\kappa}_{lk}(\widetilde{\sigma}) \sum_{j=1}^{2} \frac{\partial s}{\partial \sigma_j}(\sigma)\frac{\partial \sigma_j}{\partial \widetilde{\sigma}_k}(\widetilde{\sigma}).$$

But $\dfrac{\partial s}{\partial \sigma_1}$ and $\dfrac{\partial s}{\partial \sigma_2}$ are linearly independent, and so

$$(3.110) \qquad \sum_{k=1}^{2} \frac{\partial \sigma_k}{\partial \widetilde{\sigma}_l}\kappa_{kj}(\sigma) = \sum_{k=1}^{2} \widetilde{\kappa}_{lk}(\widetilde{\sigma})\frac{\partial \sigma_j}{\partial \widetilde{\sigma}_k}(\widetilde{\sigma}).$$

Now, if we set

$$\widetilde{K}(\widetilde{\sigma}) = [\widetilde{\kappa}_{lj}(\widetilde{\sigma})]_{l,j=1,2} \quad \text{and} \quad S(\widetilde{\sigma}) = \left[\frac{\partial \sigma_k}{\partial \widetilde{\sigma}_j}(\widetilde{\sigma})\right]_{j,k=1,2},$$

then we can write (3.110) as

$$S(\widetilde{\sigma})K(\sigma) = \widetilde{K}(\widetilde{\sigma})S(\widetilde{\sigma}).$$

However, since $\widetilde{K} = SKS^{-1}$, the eigenvalues of $K(\sigma)$ and $\widetilde{K}(\widetilde{\sigma})$ must agree. So, what we have shown is that the eigenvalues of $K(\sigma)$ in (3.105) do not depend on the way we express $\mathcal{C}_\varphi(x)$.

We would now like to fix σ at some point σ^0. In this regard, let κ be one of the eigenvalues of K, and we denote the left eigenvector of K by the row vector $\boldsymbol{a} = (a_1, a_2)$. In other words, we have assumed that $\boldsymbol{a}K = \kappa \boldsymbol{a}$. With the above, from (3.105), the following is true:

$$(3.111) \quad \left(a_1 \frac{\partial}{\partial \sigma_1} + a_2 \frac{\partial}{\partial \sigma_2}\right) \nu \bigg|_{\sigma=\sigma^0} = \kappa \left(a_1 \frac{\partial}{\partial \sigma_1} + a_2 \frac{\partial}{\partial \sigma_2}\right) s \bigg|_{\sigma=\sigma^0}.$$

The above formula shows that the way that the normal vector ν changes at $\sigma = \sigma^0$ in the direction $a_1 \partial_{\sigma_1} + a_2 \partial_{\sigma_2}$ keeps the direction $a_1 \partial_{\sigma_1} + a_2 \partial_{\sigma_2}$ while the size becomes proportional to κ.

Now, at $\sigma = \sigma_0$ we suppose that K has two distinct eigenvalues κ_1 and κ_2. Also, we can assume that the row vectors $\boldsymbol{a}_1 = (a_{11}, a_{12})$ and $\boldsymbol{a}_2 = (a_{21}, a_{22})$ are left eigenvectors that correspond, respectively, to κ_1 and κ_2. That is, we suppose that

$$(3.112) \qquad \boldsymbol{a}_l K = \kappa_l \boldsymbol{a}_l \qquad (l = 1, 2).$$

If we set $\boldsymbol{b}_l = \boldsymbol{a}_l T^{1/2}$, then by applying the operator $\boldsymbol{b}_l$ from the left in (3.107) and subsequently using (3.112), we get

$$\kappa_1 \boldsymbol{b}_l = \boldsymbol{b}_l(T^{-1/2}NT^{-1/2}) \qquad (l = 1, 2).$$

In other words, $\boldsymbol{b}_l$ ($l = 1, 2$) are the left eigenvectors corresponding to the distinct eigenvalues of the symmetric matrix $T^{-1/2}NT^{-1/2}$. Therefore, $\boldsymbol{b}_1$ and $\boldsymbol{b}_2$ are orthogonal in $\mathbb{R}^2$; i.e.,

$$0 = \boldsymbol{b}_1{}^t\boldsymbol{b}_2' = \boldsymbol{a}_1 T^t \boldsymbol{a}_2.$$

However, from the definition of T we can rewrite the above formula as

$$(3.113) \qquad \left(a_{11} \frac{\partial s}{\partial \sigma_1} + a_{12} \frac{\partial s}{\partial \sigma_2}\right) \cdot \left(a_{21} \frac{\partial s}{\partial \sigma_1} + a_{22} \frac{\partial s}{\partial \sigma_2}\right) = 0.$$

Finally, we consider the case when both the eigenvalues are equal. In this case, from (3.107), any element of $\mathbb{R}^2$ is an eigenvector. Therefore, we can choose two linearly independent vectors $\boldsymbol{a}_1$ and $\boldsymbol{a}_2$ in such a way that they satisfy (3.113).

Collecting the above together leads to the next theorem.

THEOREM 3.13. *For each fixed* $\sigma = \sigma^0$, *there exists a pair* $\boldsymbol{a}_1$ *and* $\boldsymbol{a}_2$ *of vectors such that each vector* $\boldsymbol{a}_l$ *($l = 1, 2$) satisfies (3.111), and, moreover, the two vectors satisfy (3.113).*

DEFINITION 3.14. We say that the eigenvalues $\kappa_1(\sigma)$ and $\kappa_2(\sigma)$ of $K(\sigma)$ are the *principal curvature* of the surface at $s(\sigma)$. Further, we call the element of the tangent plane $a_1 \dfrac{\partial s}{\partial \sigma_1} + a_2 \dfrac{\partial s}{\partial \sigma_2}$, determined by the eigenvalues corresponding to $K(\sigma)$, the *direction of principal curvature*.

Also, the product of the principal curvature, $\kappa_1(\sigma)\kappa_2(\sigma)$, is called the *Gaussian curvature,* and the average of the principal curvature, $\frac{1}{2}(\kappa_1(\sigma) + \kappa_2(\sigma))$, is called the *mean curvature.*

NOTE. From the above discussion, since the eigenvalues of $K(\sigma)$ do not depend on how we express the surface, the Gaussian curvature and the mean curvature are quantities that show properties of the surface, itself, no matter how we describe the surface. Moreover, if we select a unit normal vector $\nu(\sigma)$, then $-\nu(\sigma)$ is also a unit normal vector, and the elements of principal curvature corresponding to this unit normal vector also change sign. Therefore, if we are considering the principal curvature and the mean curvature, we need to be precise on which side we place the unit normal vector.

(d) The change of variable and the Laplacian.

In this section, we investigate how the Laplacian changes under a change of variable. To begin with, assume the mapping

$$(3.114) \qquad V \ni y \longrightarrow x \in U$$

from a open set V in $\mathbb{R}^3$ to some other open set U in $\mathbb{R}^3$ is smooth and that the inverse mapping exists and is also smooth. In addition, we denote the mapping (3.114) by

$$x(y) = (x_1(y_1, y_2, y_3), x_2(y_1, y_2, y_3), x_3(y_1, y_2, y_3))$$

and the inverse mapping by

$$y(x) = (y_1(x_1, x_2, x_3), y_2(x_1, x_2, x_3), y_3(x_1, x_2, x_3)).$$

Since the inverse function of $x(y)$ is $y(x)$, if we define $\dfrac{\partial x}{\partial y}$ and $\dfrac{\partial y}{\partial x}$, respectively, by

$$\frac{\partial x}{\partial y} = \left(\frac{\partial x_j}{\partial y_l}\right)_{j,l=1,2,3} \quad \text{and} \quad \frac{\partial y}{\partial x} = \left(\frac{\partial y_p}{\partial x_q}\right)_{p,q=1,2,3},$$

then we find that

$$(3.115) \qquad \frac{\partial x}{\partial y}\frac{\partial y}{\partial x} = \left(\sum_{l=1}^{3} \frac{\partial x_j}{\partial y_l}\frac{\partial y_l}{\partial x_q}\right)_{j,q=1,2,3} = (\delta_{jq})_{j,q=1,2,3} = E_3,$$

where E_3 is the unit matrix of order 3. Also, from the definition, we would like to note that

$$(3.116) \qquad {}^t\left(\frac{\partial x}{\partial y}\right)\left(\frac{\partial x}{\partial y}\right) = \left(\sum_{p=1}^{3} \frac{\partial x_p}{\partial y_j}\frac{\partial x_p}{\partial y_l}\right)_{j,l=1,2,3}.$$

Next, we denote by $J(y)$ the Jacobian matrix with the change in variable of (3.114); i.e.,

$$J(y) = \det\left[\frac{\partial x}{\partial y}(y)\right].$$

Now, let us now suppose that

$$(3.117) \qquad\qquad J(y) > 0.$$

Then, for $j, l = 1, 2, 3$, let

$$(3.118) \qquad g_{jl}(y) = \frac{\partial x}{\partial y_j}(y)\frac{\partial x}{\partial y_l}(y) = \sum_{p=1}^{3} \frac{\partial x_p}{\partial y_j}(y)\frac{\partial x_p}{\partial y_l}(y).$$

From (3.116), $[g]_{j,l=1,2,3} = {}^t\left(\dfrac{\partial x}{\partial y}\right)\left(\dfrac{\partial x}{\partial y}\right).$

Therefore, if we set

$$(3.119) \qquad\qquad g(y) = \det(g_{jl}(y))_{j,l=1,2},$$

we have that

$$(3.120) \qquad\qquad J(y) = \sqrt{g(y)}.$$

Also, if we set

$$h^{pq}(x) = \sum_{r=1}^{3} \frac{\partial y_p}{\partial x_r}(x)\frac{\partial y_q}{\partial x_r}(x),$$

then

$$\sum_{l=1}^{3} g_{jl}(y)h^{lq}(x) = \sum_{l=1}^{3}\left(\sum_{p=1}^{3}\frac{\partial x_p}{\partial y_j}\frac{\partial x_p}{\partial y_l}\right)\left(\sum_{r=1}^{3}\frac{\partial y_l}{\partial x_r}\frac{\partial y_q}{\partial x_r}\right)$$

$$= \sum_{p=1}^{3}\sum_{r=1}^{3}\left(\frac{\partial x_p}{\partial y_j}\frac{\partial y_q}{\partial x_r}\sum_{l=1}^{3}\frac{\partial x_p}{\partial y_l}\frac{\partial y_l}{\partial x_r}\right)$$

$$= \sum_{p,r=1}^{3}\delta_{pr}\frac{\partial x_p}{\partial y_j}\frac{\partial y_q}{\partial x_r} = \sum_{p=1}^{3}\frac{\partial x_p}{\partial y_j}\frac{\partial y_q}{\partial x_p} = \delta_{jq}.$$

That is, the following relation holds:

$$(3.121) \qquad (h^{pq})_{p,q=1,2,3} = [(g_{jl})_{j,l=1,2,3}]^{-1}.$$

Now, suppose that $u(x) \in C^2(U)$; then we can define $\widetilde{u}(y) \in C^2(V)$ by

$$\widetilde{u}(y) = u(x(y)).$$

Therefore, we see that

$$u(x) = \widetilde{u}(y(x)), \qquad x \in U.$$

Similarly, for $\varphi \in C_0^2(U)$, we can define $\widetilde{\varphi} \in C_0^2(V)$. So,

$$(3.122) \qquad \frac{\partial u}{\partial x_j} = \sum_{p=1}^{3}\frac{\partial \widetilde{u}}{\partial y_p}\frac{\partial y_p}{\partial x_j}, \quad \frac{\partial \varphi}{\partial x_j} = \sum_{q=1}^{3}\frac{\partial \widetilde{\varphi}}{\partial y_q}\frac{\partial y_q}{\partial x_j}.$$

By integrating by parts, we get

$$\int_U (\Delta u)(x)\varphi(x)dx = -\int_U \sum_{j=1}^{3}\frac{\partial u}{\partial x_j}(x)\frac{\partial \varphi}{\partial x_j}(x)dx,$$

and by the change in variables of (3.114),

$$\int_U (\Delta u)(x)\varphi(x)dx = -\sum_{j=1}^{3}\int_V \frac{\partial u}{\partial x_j}(x(y))\frac{\partial \varphi}{\partial x_j}(x(y))J(y)dy.$$

Next if we use (3.120) and (3.122),

$$
\int_U (\Delta u)(x)\varphi(x)dx
$$

$$
= -\sum_{j=1}^{3} \int_V \left(\sum_{p=1}^{3} \frac{\partial \tilde{u}}{\partial y_p} \frac{\partial y_p}{\partial x_j} \right) \left(\sum_{q=1}^{3} \frac{\partial \tilde{\varphi}}{\partial y_q} \frac{\partial y_q}{\partial x_j} \right) \sqrt{g(y)}dy
$$

$$
= -\sum_{p,q=1}^{3} \int_V \left(\sum_{j=1}^{3} \frac{\partial y_p}{\partial x_j} \frac{\partial y_q}{\partial x_j} \right) \frac{\partial \tilde{u}}{\partial y_p} \frac{\partial \tilde{\varphi}}{\partial y_q} \sqrt{g(y)}dy
$$

$$
= -\sum_{p,q=1}^{3} \int_V h^{pq}(x(y)) \frac{\partial \tilde{u}}{\partial y_p}(y) \frac{\partial \tilde{\varphi}}{\partial y_q}(y) \sqrt{g(y)}dy
$$

and again integration by parts leads finally to

$$
\int_U (\Delta u)(x)\varphi(x)dx = \int_V \sum_{p,q=1}^{3} \frac{\partial}{\partial y_q} \left(\sqrt{g(y)} h^{pq}(x(y)) \frac{\partial \tilde{u}}{\partial y_p}(y) \right) \tilde{\varphi}(y)dy.
$$

On the other hand, we have that

$$
\int_U (\Delta u)(x)\varphi(x)dx = \int_V (\Delta u)(x(y))\tilde{\varphi}(y)\sqrt{g(y)}dy.
$$

Hence, from the above, for arbitrary $\tilde{\varphi} \in C_0^2(V)$, the following is true:

$$
\int_V \sum_{p,q=1}^{3} \frac{\partial}{\partial y_q} \left(\sqrt{g(y)} h^{pq}(x(y)) \frac{\partial \tilde{u}}{\partial y_p}(y) \right) \tilde{\varphi}(y)dy
$$

$$
= \int_V \sqrt{g(y)}(\Delta u)(x(y))\tilde{\varphi}(y)dy.
$$

Therefore,

$$
\sqrt{g(y)}(\Delta u)(x(y)) = \sum_{p,q=1}^{3} \frac{\partial}{\partial y_q} \left(\sqrt{g(y)} h^{pq}(x(y)) \frac{\partial \tilde{u}}{\partial y_p}(y) \right).
$$

If we denote the inverse matrix of the matrix $[g_{jl}(y)]_{j,l=1,2,3}$ by $[g^{pq}(y)]_{p,q=1,2,3}$, then from (3.121) we see that $h^{pq}(x(y)) = g^{pq}(y)$ $(p, q = 1, 2, 3)$. The above now gives us the next theorem.

THEOREM 3.15. *Suppose that the change of variable in* (3.114) *is a diffeomorphism and* (3.117) *is satisfied. Then, the following holds:*

$$\Delta u(x) = \sum_{p,q=1}^{3} \frac{1}{\sqrt{g(y)}} \frac{\partial}{\partial y_q} \left(\sqrt{g(y)} g^{pq}(y) \frac{\partial \widetilde{u}}{\partial y_p}(y) \right),$$

where $\widetilde{u}$ is given by

$$\widetilde{u}(y) = u(x(y)).$$

(e) The value of $\Delta\varphi(\mathbf{x})$.

With the above preparations, we would like now to find the value of $\Delta\varphi$ of φ that satisfies $|\nabla\varphi| = 1$. We this in mind, fix $x = x_0$, and set

$$\mathcal{C} = \{x; \varphi(x) = \varphi(x_0)\}.$$

As we noted previously, $\nabla\varphi(x)$ at each point x of $\mathcal{C}$ is a unit normal vector. In a neighbourhood of x_0, we suppose that $\mathcal{C}$ is of the form

$$\{x = s(\sigma); \ \sigma = (\sigma_1, \sigma_2) \in W\},$$

where W is a neighbourhood of the origin in $\mathbb{R}^2$ and $s(\sigma)$ satisfies $s(0) = x_0$. Next, let $\nu(\sigma) = (\nabla\varphi)(s(\sigma))$. Further, if necessary by performing a linear transformation on σ, we can assume that $\dfrac{\partial s}{\partial \sigma_1}$ and $\dfrac{\partial s}{\partial \sigma_2}$ are in the direction of the principal curvature at x_0. That is to say, the following is satisfied:

(3.123) $$\frac{\partial \nu}{\partial \sigma_j}(0) = \kappa_j \frac{\partial s}{\partial \sigma_j}(0) \quad (j = 1, 2).$$

Moreover, we can assume the following holds true:

(3.124) $$\frac{\partial s}{\partial \sigma_j}(0) \cdot \frac{\partial s}{\partial \sigma_l}(0) = \delta_{jl}.$$

Now, taking $y_1 = \sigma_1$, $y_2 = \sigma_2$ and $y_3 = l$, let

(3.125) $$x(y) = s(\sigma) + l\nu(\sigma).$$

Since

$$\frac{\partial x}{\partial y}(0) = \left[\frac{\partial s}{\partial \sigma_j}(0), \ \frac{\partial s}{\partial \sigma_l}(0), \ \nu(0) \right],$$

$\det \left[\dfrac{\partial x}{\partial y}(0) \right] = \pm 1.$ So, we choose σ_1 and σ_2 in such a way that the preceding is $+1$. By choosing neighbourhoods U of x_0 and V of $y = 0$ to be suitably small, we see that

$$V \ni y \longrightarrow x = x(y) \in U$$

is a diffeomorphism.

If $|\nabla \varphi| = 1$ holds and we use (3.99), we get

$$\frac{\partial}{\partial l} \varphi(x + l\nabla\varphi(x)) = \nabla\varphi(x + l\nabla\varphi(x)) \cdot \nabla\varphi(x) = |\nabla\varphi(x)|^2 = 1.$$

So, we obtain that

$$(3.126) \qquad \widetilde{\varphi}(y) = \varphi(s(\sigma)) + l = \varphi(x_0) + l = y_3.$$

Now, we would like to find a g_{jl} given by (3.118), for the mapping given by (3.125). Since $\dfrac{\partial x}{\partial y_3} = \nu(\sigma)$ and $\dfrac{\partial \nu}{\partial \sigma_j}(\sigma) \cdot \nu(\sigma) = 0 \; (j = 1, 2)$, we have that

$$(3.127) \qquad g_{13}(\sigma) = g_{23}(\sigma) = 0, \quad g_{33}(\sigma) = 1.$$

From Theorem 3.15,

$$(3.128) \quad \Delta\varphi(x_0) = \sum_{p,q=1}^{3} \frac{1}{\sqrt{g(y)}} \frac{\partial}{\partial y_q} \left\{ \sqrt{g(y)} g^{pq}(y) \frac{\partial \widetilde{\varphi}(y)}{\partial y_p} \right\} \bigg|_{y=0}.$$

Further from (3.126), $\dfrac{\partial \widetilde{\varphi}}{\partial y_p}$ is non-zero only in the case $p = 3$, and in this case $\dfrac{\partial \widetilde{\varphi}}{\partial y_3} = 1$.

On the other hand, using (3.127), we have

$$\text{the right-hand side of (3.128)} = \frac{1}{\sqrt{g(y)}} \frac{\partial}{\partial y_3} (\sqrt{g(y)} g^{33}(y)) \bigg|_{y=0}$$

$$= \frac{1}{\sqrt{g(0)}} \frac{\partial}{\partial y_3} (\sqrt{g(0,0,y_3)} g^{33}(0,0,y_3)) \bigg|_{y_3=0}.$$

In addition, from (3.123), (3.124) and (3.127),

$$g(0,0,l) = (1 + \kappa_1 l)^2 (1 + \kappa_2 l)^2, \quad g^{33}(y) = 1.$$

Now, if we substitute this in the above, we obtain

$$(3.129) \qquad \Delta\varphi(x_0) = \kappa_1 + \kappa_2 = 2 \times (\text{mean curvature at } x_0).$$

This now shows that Theorem 3.12 is true.

COROLLARY TO THEOREM 3.12. *Let the principal curvature at* x *of* $C_\varphi(x)$ *be* κ_1, κ_2. *Then*

$$(3.130) \qquad (\Delta\varphi)(x + l\nabla\varphi(x)) = \frac{\kappa_1}{1 + l\kappa_1} + \frac{\kappa_2}{1 + l\kappa_2}.$$

PROOF. First,

$$C_\varphi(x + l\nabla\varphi(x)) = \{y; \varphi(y) = \varphi(x) + l\}.$$

Therefore, if we set $z = x + l\nabla\varphi(x)$, we can write

$$C_\varphi(z) = \{s(\sigma) + l\nu(\sigma); \sigma \in W\}.$$

Also, we know that $\nu(\sigma)$ is a unit normal vector at $s(\sigma) + l\nu(\sigma)$.
Since

$$\frac{\partial\nu}{\partial\sigma_j}(0) = \kappa_j\frac{\partial s}{\partial\sigma_j}(0) = \frac{\kappa_j}{1 + l\kappa_j}\left.\frac{\partial(s(\sigma) + l\nu(\sigma))}{\partial\sigma_j}\right|_{\sigma=0},$$

it is easy to see that $k_j(1 + l\kappa_j)^{-1}$ $(j = 1, 2)$ is the principal curvature.
Then, if we use Theorem 3.12, we establish (3.130).

$\square$

(f) An expression of the solution of the transport equation and the behaviour of the principal term.
Since $\widetilde{\kappa}(l, y) = -(\Delta\varphi)(y + 2l\nabla\varphi(y))$ in the solution (3.47) of the transport equation, then by using (3.130), we obtain

$$\int_0^s \widetilde{\kappa}(l, y)dl = -\int_0^s \left(\frac{\kappa_1(y)}{1 + 2l\kappa_1(y)} + \frac{\kappa_2(y)}{1 + 2l\kappa_2(y)}\right) dl$$
$$= -\frac{1}{2}\log(1 + 2s\kappa_1(y))(1 + 2s\kappa_2(y)),$$

where $\kappa_j(y)$ $(j = 1, 2)$ denote the principal curvature at y of $C_\varphi(y)$.
Hence,

$$(3.131) \quad \exp\left(-\int_0^s \widetilde{\kappa}(l, y)dl\right) = \{(1 + 2s\kappa_1(y))(1 + 2s\kappa_2(y))\}^{-1/2}.$$

With regard to the right-hand side of (3.131), in the case when $\kappa_1\kappa_2 \neq 0$, we would like to mention the following.

We denote the Gaussian curvature at x of $\mathcal{C}_\varphi(x)$ by $G_\varphi(x)$. So,

$$(3.132) \qquad \{(1 + t\kappa_1)(1 + t\kappa_2)\}^{-1/2} = \left\{ \frac{\dfrac{\kappa_1}{1 + t\kappa_1}\dfrac{\kappa_2}{1 + t\kappa_2}}{\kappa_1 \kappa_2} \right\}^{1/2}$$

$$(3.133) \qquad\qquad\qquad\qquad = \{G_\varphi(y + t\nabla\varphi(y))/G_\varphi(y)\}^{1/2}.$$

Also, if we let $x = y + t\nabla\varphi(y)$ and since $\kappa_l(x) = \dfrac{\kappa_l(y)}{1 + t\kappa_l(y)}$
$(l = 1, 2)$, we have

$$(3.134) \quad \{(1 + t\kappa_1(y))(1 + t\kappa_2(y))\}^{-1/2} = \{(1 - t\kappa_1(x))(1 - t\kappa_2(x))\}^{1/2}.$$

Now, using the fact that $T(s) = t_0 + 2s$, from (3.51) we see that
$$(3.135)$$
$$v_0(t_0 + t, y + t\nabla\varphi(y)) = v_0(t_0, y)\{(1 + t\kappa_1(y))(1 + t\kappa_2(y))\}^{-1/2}.$$

In addition, suppose that U is a closed set in Ω and the following holds:
$$\operatorname{supp} v_j(t_0, \cdot) \subset U, \quad j = 0, 1, 2, \ldots, m + N.$$
So, if we apply Proposition 3.7, we get

$$(3.136) \qquad \operatorname{supp} v_j(t_0 + t, \cdot) \subset \{y + t\nabla\varphi(y); y \in U\}.$$

This formula, now, shows that the asymptotic solution u as time progresses moves directly in the direction $\nabla\varphi$ with speed 1.
    If
$$u_0(t, x; k) = e^{ik(t - \varphi(x))} v_0(t, x),$$
then by letting $x = y + l\nabla\varphi(y)$ and from the fact that $\varphi(y + l\nabla\varphi(y)) = \varphi(y) + l$, we can write

$$(3.137) \qquad \begin{cases} u_0(t_0 + t, x; k) = u_0(t_0, x - t\nabla\varphi(x); k)A(x, t), \\ A(x, t) = \{(1 - t\kappa_1(x))(1 - t\kappa_2(x))\}^{1/2}. \end{cases}$$

This shows that u_0, moving from the point y in the direction $\nabla\varphi(y)$ with speed 1, can only change in one way; namely, its size will change in accordance with the curvature of the wave front. Also, from this formula, regarded as being parameterized by t, we see that the graph of $u_0(t_0 + t, x; k)$, regarded as a function in x, is obtained from the graph of the function $u_0(t_0, x)$ in x as follows. At each point, first, translate the graph by a length t in the direction $\nabla\varphi$, then increase or decrease its height according to the state of the curvature of the wave front. So, the wave moves directly in the direction $\nabla\varphi$.

If $\nabla\varphi(x)$ is a fixed vector, i.e., there is an $\omega \in \mathbb{R}^3$ such that $|\omega| = 1$, and for this case we may write $\varphi(x) = x \cdot \omega$, then since $\kappa_1(x) = \kappa_2(x) = 0$, we have

$$u_0(t_0 + t, c; k) = u_0(t_0, x - t\omega; k),$$

and the graph of u_0 as a function of x moves in a parallel manner with time in the direction ω with speed 1.

Next, with x fixed, we look at $u_0(t_0 + t, x)$ as a function in t. From (3.135), we can write

$$u_0(t_0 + t, x) = e^{ikt}v_0(t, x - t\nabla\varphi(x))A(x, t)e^{ik(t_0 - \varphi(x))}.$$

If we set $\widetilde{v}_0(t) = v_0(t, x - t\nabla\varphi(x))A(x, t)e^{ik(t_0 - \varphi(x))}$, then

(3.138) $u_0(t_0 + t, x) = e^{ikt}\widetilde{v}_0(t).$

Now, if k is sufficiently large, then this shows that the function denoted by (3.138) has a simple oscillation with frequency $k/2\pi$ and amplitude $\widetilde{v}_0(t)$.

Gathering all this together, we can deduce the next theorem.

THEOREM 3.16. *The asymptotic solution of the wave equation given in (3.96) can be expressed as a wave with frequency $k/2\pi$ that moves directly in the direction of $\nabla\varphi$ with velocity 1.*

(g) The solution when the wave equation incorporates friction.

Suppose that

$$P = \Box + h_0(x)\frac{\partial}{\partial t},$$

where the friction coefficient, denoted by $h_0(x)$, is such that there exists a positive constant α_0 with $h_0(x) \geqslant \alpha_0$. Since the friction term is of low order, the *Eikonal* equation is not affected. So, the operator T that appears in the transport equation is

$$T = 2\frac{\partial}{\partial t} + 2\nabla\varphi \cdot \nabla + h_0(x) + \Delta\varphi.$$

Therefore, on replacing (3.137), the principal term in the asymptotic solution is

$$u_0(t_0 + t, y + t\nabla\varphi(y); k) = u_0(t_0, y; k) \exp\left(-\int_0^t h_0(y + l\nabla\varphi(y)) dl\right)$$
$$\times \{(1 + t\kappa_1)(1 + t\kappa_2)\}^{1/2}.$$

Now, since $h_0 \geqslant \alpha_0 > 0$, according to the amount of friction, the amplitude exponentially decreases with time.

(h) Reflection in the wave equation.

In the case when the space in which the wave is transmitted has a boundary, on hitting the boundary it is quite common for the wave to be reflected. So, we look at this by means of the asymptotic solution.

In the above regard, let $\Omega \in \mathbb{R}^3$ be a domain with smooth boundary Γ. Next, consider the asymptotic solution that satisfies

$$(3.139) \qquad \begin{cases} \Box u = 0 & \text{in } (0, \infty) \times \Omega, \\ Bu = 0 & \text{on } (0, \infty) \times \Gamma. \end{cases}$$

In the following we consider the case $Bu = u$, namely the case where the Dirichlet boundary condition is imposed.

To begin with, we denote by u^- the asymptotic solution that corresponds to the incidence ray, and suppose it is of the form

$$(3.140) \qquad u^-(t, x; k) = e^{ik(\varphi^-(x)-t)} v^-(t, x; k)$$

with

$$v^-(t, x; k) = \sum_{j=0}^N v_j^-(t, x)(ik)^{-j}.$$

Further, assume that the following holds:

$$(3.141) \qquad \Box u^- = O(k^{-N}).$$

Since we are considering the effect of the reflection upon the solution at the boundary, we can assume that the support of $u^-(t, \cdot\,; k)$ for small t is contained in Ω, and further with time the support of $u^-(t, \cdot\,; k)$ intersects with Γ. So, we are looking for the asymptotic solution u^+ corresponding to the reflected wave of the form

$$(3.142) \qquad u^+(t, x; k) = e^{ik(\varphi^+(x)-t)} v^+(t, x; k)$$

with

$$v^+(t,x;k) = \sum_{j=0}^{N} v_j^+(t,x)(ik)^{-j}.$$

Further, u^+ is such that

(3.143) $$\Box u^+ = O(k^{-N}) \quad \text{in } \Omega$$

and

(3.144) $$u^- + u^+ = 0 \quad \text{on } \Gamma.$$

Now, in order for (3.143) to be satisfied, it suffices that

$$|\nabla \varphi^+| = 1, \quad K^+ v_j^+ = -\Box v_{j-1}^+ \quad (j = 0, 1, 2, \ldots, N),$$

with v_{-1}^+ defined to be zero. In addition, for (3.144) to be true, it is sufficient that the following hold:

(3.145) $$\varphi^+(x) = \varphi^-(x) \quad \text{on } \Gamma$$

and

(3.146) $$v_j^+(t,x) = -v_j^-(t,x) \quad \text{on } \Gamma \quad (j = 0, 1, 2, \ldots, N).$$

To begin, we concern ourselves with the φ^+ that satisfies (3.145). In the neighbourhood of $x^0 \in \Gamma$ suppose Γ is expressed as

$$x = s(\sigma), \quad \sigma \in W.$$

Since from (3.145) we see that $\varphi^+(s(\sigma)) = \varphi^-(s(\sigma))$ $(\sigma \in W)$, by partially differentiating with respect to σ_j $(j = 1, 2)$ we have that

$$\sum_{l=1}^{3} \frac{\partial \varphi^+}{\partial x_l} \frac{\partial s_l}{\partial \sigma_j} = \sum_{l=1}^{3} \frac{\partial \varphi^-}{\partial x_l} \frac{\partial s_l}{\partial \sigma_j} \quad (j = 1, 2);$$

i.e.,

(3.147) $$\nabla \varphi^+ \cdot \frac{\partial s}{\partial \sigma_j} = \nabla \varphi^- \cdot \frac{\partial s}{\partial \sigma_j} \quad (j = 1, 2).$$

Next, suppose that $\nu(\sigma)$ is a unit normal vector at $s(\sigma) \in \Gamma$ in Ω. So, if we set

$$\nabla \varphi^\pm = \nabla \varphi^\pm - (\nabla \varphi^\pm \cdot \nu)\nu + (\nabla \varphi^\pm \cdot \nu)\nu,$$

and use that $\nu \cdot \dfrac{\partial s}{\partial \sigma_j} = 0$, from (3.147) we obtain that

$$(3.148) \quad (\nabla\varphi^+ - (\nabla\varphi^+ \cdot \nu)\nu) \cdot \frac{\partial s}{\partial \sigma_j} = (\nabla\varphi^- - \nabla\varphi^- \cdot \nu) \cdot \frac{\partial s}{\partial \sigma_j} \quad (j = 1, 2).$$

Now, since $\dfrac{\partial s}{\partial \sigma_j}$ $(j = 1, 2)$ form a basis for $T_s\Gamma$ and, moreover, since $(\nabla\varphi^\pm - (\nabla\varphi^\pm \cdot \nu)\nu)$ is an element of $T_s\Gamma$, then with the help of (3.148), we see that

$$(3.149) \qquad \nabla\varphi^+ - (\nabla\varphi^+ \cdot \nu)\nu = \nabla\varphi^- - (\nabla\varphi^- \cdot \nu)\nu.$$

From this and further that $|\nabla\varphi^+| = |\nabla\varphi^-|$, the size of the components orthogonal to $T_s\Gamma$ of $\nabla\varphi^\pm$ is the same for both, i.e., $|(\nabla\varphi^+ \cdot \nu)| = |(\nabla\varphi^- \cdot \nu)|$.

So, we assume that $\nabla\varphi^+ \neq \nabla\varphi^-$, if we take that (3.149) already holds, then we must have that

$$(3.150) \qquad \nabla\varphi^+ \cdot \nu = -\nabla\varphi^- \cdot \nu.$$

From the above, we get that

$$
\begin{aligned}
(3.151) \qquad \nabla\varphi^+ &= \nabla\varphi^+ - (\nabla\varphi^+ \cdot \nu)\nu + (\nabla\varphi^+ \cdot \nu)\nu \\
&= \nabla\varphi^- - (\nabla\varphi^- \cdot \nu)\nu - (\nabla\varphi^- \cdot \nu)\nu \\
&= \nabla\varphi^- - 2(\nabla\varphi^- \cdot \nu)\nu.
\end{aligned}
$$

The upshot of (3.150) and (3.151) is the following theorem.

THEOREM 3.17. *For the solutions φ^+ and φ^- of the Eikonal equation that satisfies (3.145) if we have $\nabla\varphi^+ \neq \nabla\varphi^-$, then for every point $x \in \Gamma$, we may regard $\nabla\varphi^-$ as the ray of incidence and $\nabla\varphi^+$ as the ray of reflection. Further, φ^+ and φ^- satisfy the law of reflection in geometric optics.*

The above may be interpreted as follows, the ray of reflection is on the plane that is determined by the normal vector to the boundary at the point of reflection, and further, it is, with respect to the normal vector, on the opposite side to the ray of incidence. Finally, the angle that the ray of reflection makes with the normal is equal to the angle the ray of incidence makes with the normal.

Since we understand the behaviour of the phase function φ^+ of the asymptotic solution for the reflected wave, the construction of u^+ is easy if we follow the methods of the previous sections. That is, it suffices to solve the transport equation with the conditions in (3.146).

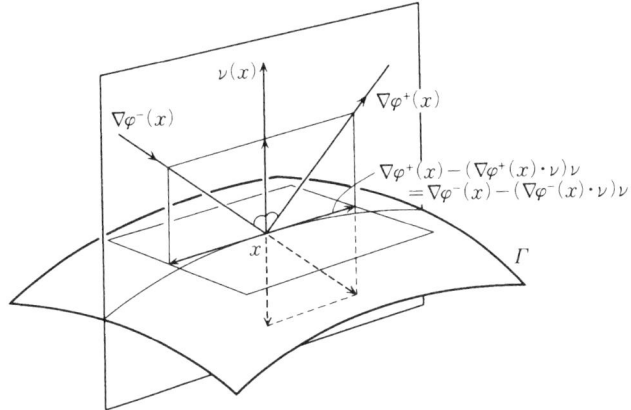

FIGURE 3.4

For the solutions u^+ and u^- formed in this way, we denote $u = u^- + u^+$; then u satisfies (3.139) asymptotically. So, we investigate the behaviour of this u.

To start with, we assume that at $t = 0$, the support of $u(t, \cdot\,; k)$ is contained in U. As the support is transmitted with time in the direction of $\nabla\varphi^-$ and velocity 1, it will hit the boundary which causes a subsequent change in the phase by only $1/2$ (corresponding to the fact that the sign of the amplitude function changes) and is reflected. The direction of the reflection at each point on the boundary is determined by the laws of geometric optics.

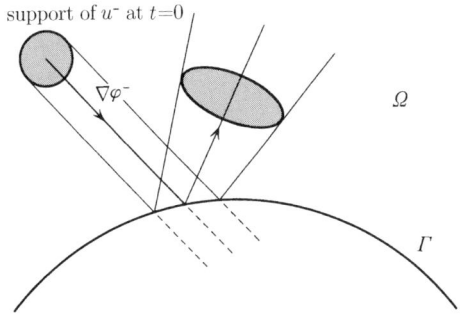

FIGURE 3.5

If the wave front of the incident wave is flat, the size of the amplitude function undergoes no change, but, as in Figure 3.5, if Γ is curved, the wave front of the reflected wave bends correspondingly, and so the amplitude changes.

After only time t, the oscillating position, i.e., $\text{supp}\, u(t, \cdot\,; k)$, is contained in the set constructed as follows. From every point of U there emerges a straight line in the direction of $\nabla\varphi^-$, if it hits Γ, then in accordance with geometric optics, it will be reflected as a polygonal line. The set, itself, is the points on these reflected lines whose length is time t from the starting point.

(i) Refraction.

Suppose that on either side of the boundary surface Γ the characteristics of the media are different. So, we denote one medium by Ω_1 and the other by Ω_2. For example, suppose Ω_1 consists of air and Ω_2 consists of water. The natural question to ask is, "As light passes from the medium containing air into the water how does the direction of the propagation wave change at the boundary interface?" In this section, we look at such phenomena.

With this in mind, suppose λ_1 is the propagation speed in Ω_1 and λ_2 is the propagation speed in Ω_2. Therefore, if we set

$$P_1 = \frac{\partial^2}{\partial t^2} - \lambda_1^2 \Delta, \quad P_2 = \frac{\partial^2}{\partial t^2} - \lambda_2^2 \Delta,$$

the problem is to study the behaviour of the functions $u^{(l)}$ $(l = 1, 2)$ defined on the same domains in $(0, \infty) \times \overline{\Omega_l}$ $(l = 1, 2)$ and which satisfy

(3.152)
$$\begin{cases} P_1 u^{(1)} = 0 & \text{in } (0, \infty) \times \Omega_1, \\ P_2 u^{(2)} = 0 & \text{in } (0, \infty) \times \Omega_2, \\ u^{(1)} \text{ and } u^{(2)} \text{ satisfy the interface condition on } \Gamma. \end{cases}$$

First, determine the interface condition. So, let $\nu(x)$ be the unit normal vector of Γ at $x \in \Gamma$ directed towards the Ω_1-side. Further, we denote the normal differential $\sum\limits_{j=1}^{3} \nu_j u_{x_j}$ by $\dfrac{\partial u}{\partial \nu}$. Also, we write $u(t, x + 0\nu(x))$ and $u(t, x - 0\nu(x))$ for the boundary values at Γ from the Ω_1-side and the Ω_2-side, respectively. To be precise, for $x \in \Gamma$, we set

$$u(t, x \pm 0\nu(x)) = \lim_{\varepsilon \to +0} u(t, x \pm \varepsilon \nu(x)).$$

Now, suppose that the boundary condition is given as follows:

(3.153)
$$\begin{cases} u^{(1)}(t, x + 0\nu(x)) = au^{(2)}(t, x - 0\nu(x)), \quad x \in \Gamma, \\ \dfrac{\partial u^{(1)}}{\partial \nu}(t, x + 0\nu(x)) = b\dfrac{\partial u^{(2)}}{\partial \nu}(t, x - 0\nu(x)), \quad x \in \Gamma, \end{cases}$$

where a and b are positive constants.

In order to simplify the notation for our subsequent discussions, we will write $u(t, x)$ $(x \in \Gamma)$ for the continuous function u on $\mathbb{R} \times \overline{\Omega_1}$ instead of $u(t, x + 0\nu(x))$ and $\mathbb{R} \times \overline{\Omega_2}$ instead of $u(t, x - 0\nu(x))$.

Now, at $t = 0$ we will take as the incidence wave the asymptotic wave u^- which is transmitted towards the interface and has support in the set U in Ω_1, disjoint from Γ. We will assume that u^- has the form

(3.154)
$$u^-(t, x; k) = e^{ik(t-\varphi^-(x))}v^-(t, x; k)$$

with

(3.155)
$$v^{-1}(t, x; k) = \sum_{j=0}^{J} v_j^-(t, x)(ik)^{-j}.$$

For $t - \varphi^-(x)$ to satisfy the *Eikonal* equation, we must have that

(3.156)
$$|\nabla \varphi^-|^2 = \lambda_1^{-2}.$$

By considering the case for which u^- moves in the direction of $\nabla \varphi^-$ and hits Γ, we can assume that $\nabla \varphi^-(x) \cdot \nu(x) < 0$.

For the incidence wave u^- since we are investigating reflection and refraction, we seek a $u^{(1)}$ on the Ω_1-side with the form

(3.157)
$$u^{(1)}(t, x; k) = u^-(t, x; k) + u^+(t, x; k),$$
$$\begin{cases} u^+(t, x; k) = e^{ik(t-\varphi^+(x))}v^+(t, x; k), \\ v^+(t, x; k) = \sum_{j=0}^{N}(ik)^{-j}v_j^+(t, x), \end{cases}$$

and, if we set $u^{(2)} = u^d$, on the Ω_2-side with the form

(3.158)
$$\begin{cases} u^d(t, x; k) = e^{ik(t-\phi(x))}v^d(t, x; k), \\ v^d(t, x; k) = \sum_{j=0}^{N}(ik)^{-j}v_j^d(t, x). \end{cases}$$

In order for u^+ on Ω_1 and u^d on Ω_2 to satisfy, respectively, the *Eikonal* equation, the following must hold:

(3.159)
$$|\nabla\varphi^+(x)|^2 = \lambda_1^{-2},$$

(3.160)
$$|\nabla\phi(x)|^2 = \lambda_2^{-2}.$$

Now, for the above $u^{(1)}$ and $u^{(2)}$ to satisfy the interface conditions, we desire that

(3.161)
$$\varphi^-(x) = \varphi^+(x) \quad \text{on } \Gamma,$$

(3.162)
$$\varphi^-(x) = \phi(x) \quad \text{on } \Gamma.$$

So, first we consider (3.162), and at a neighbourhood of $x^0 \in \Gamma$ we assume that Γ is expressed as $x = s(\sigma)$ with $\sigma \in W$. From (3.162), we obtain
$$\varphi^-(s(\sigma)) = \phi(s(\sigma)), \quad \forall \sigma \in W.$$

Hence, by partially differentiating with respect to σ_j $(j = 1, 2)$, we obtain

(3.163) $\quad (\nabla\varphi^-)(s(\sigma)) \cdot \dfrac{\partial s}{\partial \sigma_j}(\sigma) = (\nabla\phi)(s(\sigma)) \cdot \dfrac{\partial s}{\partial \sigma_j}(\sigma) \quad (j = 1, 2).$

But, $\nabla\varphi^- - (\nabla\varphi^- \cdot \nu)\nu$ and $\nabla\phi - (\nabla\phi \cdot \nu)\nu$ are in $T_s\Gamma$, so (3.163) shows that

(3.164)
$$\nabla\varphi^- - (\nabla\varphi^- \cdot \nu)\nu = \nabla\phi - (\nabla\phi \cdot \nu)\nu.$$

Now, if we assume that (3.164) holds for $\nabla\phi \in \mathbb{R}^3$ and $|\nabla\phi| = \lambda_2^{-1}$, clearly, we must have that

(3.165)
$$|\nabla\varphi^- - (\nabla\varphi^- \cdot \nu)\nu| \leqslant \lambda_2^{-1}.$$

Next, for $x \in \Gamma$, let the angle that $\nu(x)$ makes with $-\nabla\varphi^-(x)$ be θ_i and the angle that $-\nu(x)$ makes with $\nabla\phi(x)$ be θ_d. But, $|\nabla\varphi^- - (\nabla\varphi^- \cdot \nu)\nu| = \lambda_1^{-1}\sin\theta_i$, so, (3.165) now becomes

(3.166)
$$\sin\theta_i \leqslant \frac{\lambda_1}{\lambda_2}.$$

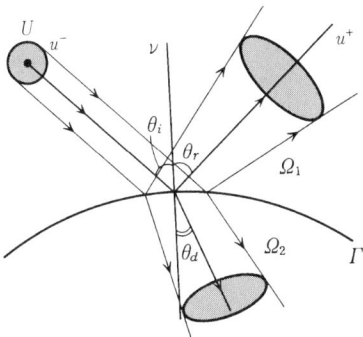

FIGURE 3.6

If $\lambda_1 > \lambda_2$, then (3.166) does not give any restrictions on θ_i, however, if $\lambda_1 < \lambda_2$ we do, in fact, get some restrictions on the angle of incidence, θ_i. If (3.166) holds, we may assume the existence of ϕ that satisfies (3.162).

Then, at each point of Γ,

$$\frac{|\nabla\varphi^- - (\nabla\varphi^- \cdot \nu)\nu|}{|\nabla\varphi^-|} = \frac{\lambda_1}{\lambda_2} \cdot \frac{|\nabla\phi - (\nabla\phi \cdot \nu)\nu|}{|\nabla\phi|};$$

i.e., the following holds:

$$(3.167) \qquad \frac{\sin\theta_i}{\sin\theta_d} = \frac{\lambda_1}{\lambda_2}.$$

The above formula is commonly known as *Snell's law* and gives us the relationship between the angle of incidence and the angle of refraction.

On the other hand, since

$$|\nabla\varphi^- - (\nabla\varphi^- \cdot \nu)\nu|^2 + (\nabla\varphi^- \cdot \nu)^2 = \lambda_1^{-2},$$
$$|\nabla\phi - (\nabla\phi \cdot \nu)\nu|^2 + (\nabla\phi \cdot \nu)^2 = \lambda_2^{-2},$$

then by using (3.164), we obtain that

$$(3.168) \qquad (\nabla\phi \cdot \nu)^2 = \lambda_2^{-2} - \lambda_1^{-2} + (\nabla\varphi^- \cdot \nu)^2.$$

Also, since $\nabla\phi \cdot \nu < 0$, we have

$$(3.169) \qquad \frac{\partial\phi}{\partial\nu} = -\sqrt{\lambda_2^{-2} - \lambda_1^{-2} + (\nabla\varphi^- \cdot \nu)^2}.$$

From the above formula and (3.164), it follows that

$$
\nabla\phi = \nabla\varphi^- - \left\{ (\nabla\varphi^- \cdot \nu) + \sqrt{\lambda_2^{-2} - \lambda_1^{-2} + |\nabla\varphi^- \cdot \nu|^2} \right\} \nu
$$

(3.170)

$$
= \nabla\varphi^- - \frac{\lambda_2^{-2} - \lambda_1^{-2}}{\sqrt{\lambda_2^{-2} - \lambda_1^{-2} + (\nabla\varphi^- \cdot \nu)^2} - |\nabla\varphi^- \cdot \nu|} \nu.
$$

For the φ^+ that satisfies (3.161), this was investigated when we considered reflection in the previous section. Based on the above preparations, we are looking for an amplitude function that satisfies the interface condition. If the respective phase functions (3.161) and (3.162) are fulfilled, then the interface condition (3.153) is the same as the following:

(3.171)
$$
\begin{cases}
v^-(t, x + 0\nu; k) + v^+(t, x + 0\nu; k) = av^d(t, x - 0\nu; k), \\
\left(ik\dfrac{\partial\varphi^-}{\partial\nu} v^- + ik\dfrac{\partial\varphi^+}{\partial\nu} v^+ + \dfrac{\partial\varphi^-}{\partial\nu} + \dfrac{\partial\varphi^+}{\partial\nu} \right)(t, x + 0\nu; k) \\
\quad = b\left(ik\dfrac{\partial\phi}{\partial\nu} v^d + \dfrac{\partial v^d}{\partial\nu} \right)(t, x - 0\nu; k).
\end{cases}
$$

By substituting (3.155), (3.157) and (3.158) for $v^\pm$ and v^d, respectively, and then comparing the coefficients of the powers of ik, we must have the following relations between $v_0^\pm$ and v_0^d on Γ:

(3.172)
$$
\begin{cases}
v_0^- + v_0^+ = av_0^d, \\
\dfrac{\partial\varphi^-}{\partial\nu} v_0^- + \dfrac{\partial\varphi^+}{\partial\nu} v_0^+ = b\dfrac{\partial\phi}{\partial\nu} v_0^d.
\end{cases}
$$

Now, noting that $\dfrac{\partial\varphi^+}{\partial\nu} = -\dfrac{\partial\varphi^-}{\partial\nu} > 0$, if we assume that v_0^- is known, the following holds from (3.172):

(3.173)
$$
\begin{cases}
v_0^+ = -\left(b\dfrac{\partial\phi}{\partial\nu} + a\dfrac{\partial\varphi^-}{\partial\nu} \right)^{-1} \left(b\dfrac{\partial\phi}{\partial\nu} - a\dfrac{\partial\varphi^-}{\partial\nu} \right) v_0^-, \\
v_0^d = 2\left(b\dfrac{\partial\phi}{\partial\nu} + a\dfrac{\partial\varphi^-}{\partial\nu} \right)^- \dfrac{\partial\varphi^-}{\partial\nu} v_0^-.
\end{cases}
$$

Continuing in the above fashion, for $v_j^-(t, x)$ $(x \in \Gamma)$ we can determine successively the values of v_j^+ and v_j^d $(j = 1, 2, \dots, N)$ on Γ.

In this way, if we determine the values of v^+ and v^d on Γ, then we can successively solve the transport equation, and so determine

v^+ on Ω_1 and v^d on Ω_2. For u^+ and u^d so formed, if, as previously explained, u is defined by

$$u(t,x;k) = \begin{cases} u^-(t,x;k) + u^+(t,x;k), & x \in \Omega_1, \\ u^d(t,x;k), & x \in \Omega_2, \end{cases}$$

then u satisfies (3.152) with an error of $O(k^{-N})$.

Let us now look at the principal parts of $u^{\pm}$ and u^d. If k is large, then we note that the behaviour of the solution can be essentially decided by this principal part.

The case of λ_1 and λ_2 being reasonably close.

Since $\dfrac{\partial \phi}{\partial \nu}$ is sufficiently close to $\dfrac{\partial \varphi^-}{\partial \nu}$, we see that

$$v_0^+ \doteqdot -\frac{a-b}{a+b} v_0^-, \qquad v_0^d \doteqdot \frac{2}{a+b} v_0^- \quad \text{on } \Gamma.$$

If a and b are close, then the amount that is reflected is small, with the majority being refracted into Ω_2. The strength of the reflection can be determined from the coefficients a and b.

The case of $\lambda_1 > \lambda_2$ and θ_i is close to $\pi/2$.

This case corresponds to the speed of propagation through Ω_1 being larger than that through Ω_2 (for instance, when a light ray passes from air into water), and the angle of incidence is close to being a right angle. So, it is the case of $\dfrac{\partial \varphi^-}{\partial \nu} \doteqdot 0$.

Since $\dfrac{\partial \phi}{\partial \nu} \doteqdot -(\lambda_2^{-2} - \lambda_1^{-2})^{-1/2}$ from (3.169), we have that

$$v_0^+ \doteqdot -v_0^-, \quad v^d \doteqdot 0 \qquad \text{on } \Gamma.$$

What the above is saying is that if the incident light ray enters by skimming along the boundary surface, then almost certainly it is reflected by the boundary and does not enter the Ω_2-side.

Total reflection, the case of $\lambda_1 < \lambda_2$.

To find v^d we must find a ϕ that satisfies $(\nabla \phi)^2 = \lambda_2^{-2}$ and (3.164) for φ^-. Now, if $\lambda_2 > \lambda_1$ and we restrict ourselves to real-valued functions, then there may not be functions that satisfy these two conditions simultaneously. In fact, for the angle of incidence θ_i, the necessary condition for the existence of a real-valued function ϕ is (3.166). Also, similarly, (3.168) cannot hold, but if

$$(3.174) \qquad (\nabla \varphi^- \cdot \nu)^2 < \lambda_1^{-2} - \lambda_2^{-2},$$

then (3.168) is never true when ϕ is a real-valued function. So, the case of (3.174) requires that we search for a complex-valued function ϕ that satisfies (3.160) and (3.162).

With the above in mind, suppose that there exists an $\alpha > 0$ such that

$$(3.175) \qquad \lambda_1^{-2} - \lambda_2^{-2} - (\nabla \varphi^- \cdot \nu)^2 \geqslant \alpha^2.$$

So, we are seeking a ϕ of the form
(3.176)
$$\phi(s - l\nu(s)) = \phi_0(s) + l\phi_1(s) + \frac{l^2}{2!}\phi_2(s) + \cdots + \frac{l^m}{m!}\phi_m(s) \quad (s \in \Gamma, \ l \geqslant 0).$$

Now, in order for (3.162) to hold we must have that

$$\phi_0(s) = \varphi^-(s) \quad (s \in \Gamma).$$

Using the notation of (3.125), we set $y = (\sigma_1, \sigma_2, l)$ and $\widetilde{\phi}(y) = \phi(s(\sigma) - l\nu(s))$. Then,

$$(\nabla\phi)^2 = \sum_{p,q=1}^{3} g^{pq}(y) \frac{\partial\widetilde{\phi}}{\partial y_p} \frac{\partial\widetilde{\phi}}{\partial y_q}.$$

However, since

$$(g^{pq})_{p,q=1,2,3} = [(g_{jl})_{j,l=1,2,3}]^{-1} \quad \text{and} \quad g_{jl} = \frac{\partial x}{\partial y_j} \cdot \frac{\partial x}{\partial y_l},$$

from $g_{13} = g_{23} = 0$ and $g_{33} = 1$, we have that $g^{13} = g^{23} = 0$ and $g^{33} = 1$. Hence,

$$\begin{aligned}
(\nabla\phi)^2 &= \sum_{p,q=1}^{2} g^{pq}(y) \frac{\partial\widetilde{\phi}}{\partial y_p} \frac{\partial\widetilde{\phi}}{\partial y_q} + \left(\frac{\partial\widetilde{\phi}}{\partial l}\right)^2 \\
&= \left(\phi_1 + l\phi_2 + \cdots + \frac{l^{m-1}}{(m-1)!}\phi_m\right)^2 \\
&\quad + \sum_{p,q=1}^{2} g^{pq}(\sigma, l) \left(\frac{\partial\phi_0}{\partial\sigma_p} + l\frac{\partial\phi_1}{\partial\sigma_p} + \cdots + \frac{l^m}{m!}\frac{\partial\phi_m}{\partial\sigma_p}\right) \\
&\quad \times \left(\frac{\partial\phi_0}{\partial\sigma_q} + l\frac{\partial\phi_1}{\partial\sigma_q} + \cdots + \frac{l^m}{m!}\frac{\partial\phi_m}{\partial\sigma_q}\right).
\end{aligned}$$

But, since $g^{pq}(\sigma, l)$ is smooth in σ and l, we can construct a Taylor expansion in l. Let us determine $\phi_1, \phi_1, \ldots, \phi_m$ so that

(3.177) $$(\nabla\phi)^2 = \lambda_2^{-2} + O(l^m).$$

Next, we will set to be equal, respectively, the coefficients of equal powers of l in both sides of the expansion in (3.176). Now, since

$$\sum_{p,q=1}^{2} g^{pq}(\sigma, 0) \frac{\partial\phi}{\partial\sigma_p}(\sigma, 0) \frac{\partial\phi}{\partial\sigma_q}(\sigma, 0)$$
$$= (\nabla\phi - (\nabla\phi \cdot \nu)\nu)^2 = \lambda_1^{-2} - (\nabla\varphi^- \cdot \nu)^2,$$

and comparing the coefficient of l^0, we see that

$$\lambda_2^{-2} = \lambda_1^{-2} - (\nabla\varphi^- \cdot \nu)^2 + \phi_1^2.$$

If we choose $\operatorname{Im}\phi_1 < 0$, we can deduce that

(3.178) $$\phi_1 = -i\{\lambda_1^{-2} - \lambda_2^{-2} - (\nabla\varphi^- \cdot \nu)^2\}^{1/2}.$$

Next, by comparing the coefficient of l^1, we obtain that

$$2\phi_1\phi_2 + 2\sum_{p,q=1}^{2}\left\{ g^{pq}(\sigma, 0)\frac{\partial\phi_0}{\partial\sigma_p}\frac{\partial\phi_1}{\partial\sigma_q} + g_1^{pq}(\sigma, 0)\frac{\partial\phi_1}{\partial\sigma_p}\frac{\partial\phi_0}{\partial\sigma_q}\right\} = 0,$$

where $g_1^{pq}(\sigma, 0) = \partial_l g^{pq}(\sigma, l)|_{l=0}$.

From the assumption in (3.175) since $|\phi_1| \geqslant c$, we can determine ϕ_2 from this relation. Proceeding in this way, we will eventually evaluate $\phi_3, \phi_4, \ldots, \phi_m$.

With the ϕ_j $(j = 0, 1, 2, \ldots, m)$ determined as above, we have that ϕ, as defined by (3.176), satisfies (3.177).

Now, for this ϕ we would like to determine v^+ and v^d in such a way that they satisfy (3.171). The value on Γ of v_0^+ and v_0^d is given by (3.173), but since $\left.\dfrac{\partial\phi}{\partial\nu}\right|_\Gamma = \phi_1$ and ϕ_1 is purely imaginary, we see that

(3.179) $$|v_0^+| = |v_0^-| \qquad \text{on } \Gamma.$$

When v_0^d is given on Γ, we will construct a v_0^d that satisfies

$$T_\phi v_0^d = O(l^m)$$

and is of the form

$$v_0^d(t, s - l\nu(s)) = v_{00}^d(t, s) + lv_{01}^d(t, s) + \cdots + \frac{l^m}{m!}v_{0m}^d(t, s).$$

This procedure is the same as that used in the construction of ϕ. By means of this, we can step by step construct v_j^+ and v_j^d ($j = 0, 1, 2, \ldots, N$).

So, from (3.178), there exists an $l_0 > 0$ such that

$$(3.180) \qquad -\operatorname{Im}\phi(s - l\nu(s)) \geqslant \frac{c}{2}l \qquad (0 \leqslant l \leqslant 2l_0).$$

We now take $\chi(l) \in C^\infty(\mathbb{R})$ such that

$$\chi(l) = \begin{cases} 1 & (l \leqslant l_0), \\ 0 & (l \geqslant 2l_0). \end{cases}$$

By setting $x = s - l\nu(s)$, we define u^d by

$$u^d(t, x; k) = e^{ik(t - \phi(x))}v^d(t, x; k)\chi(l).$$

Since

$$P_2 u^d = e^{ik(t - \phi(x))}\{(ik)^2(1 - \lambda_2^2(\nabla\phi)^2)v^d + ikT_\phi v^d + P_2 v^d\},$$

then for $0 \leqslant l \leqslant l_0$ we have that

$$|P_2 u^d| \leqslant Ck^2 e^{k \operatorname{Im}\phi}l^m,$$

while for $l_0 \leqslant l$,

$$|P_2 u^d| \leqslant Ck^2 e^{k \operatorname{Im}\phi}.$$

But, if we now use (3.180) we see that

$$k^2 e^{-k \operatorname{Im}\phi}l^m \leqslant k^2 l^m e^{-1/2kcl} \leqslant Ck^{-m+2} \qquad (0 \leqslant l \leqslant l_0)$$

and

$$k^2 e^{-k \operatorname{Im}\phi} \leqslant k^2 l^m e^{-1/2kcl_0} \qquad (2l_0 \geqslant l \geqslant l_0).$$

Due to the method of construction of v^+ and v^d, the interface condition is satisfied up to a degree of $O(k^{-m+2})$.

From the above,

$$(3.181) \qquad u(t, x; k) = \begin{cases} u^-(t, x; k) + u^+(t, x; k), & x \in \Omega_1, \\ u^d(t, x; k), & x \in \Omega_2, \end{cases}$$

is the asymptotic solution of the interface problem (3.152).

Let us now investigate, via the principal part, the behaviour of the asymptotic solution u given by (3.181). From (3.179), we see that the absolute value of the amplitude of the reflected wave v^+ is the same as that of the incident wave u^-. If the difference in the phase of the waves is negligible, then this is the same as reflection at a boundary (for example due to a mirror) considered in section (h). For u^d at the point that is only a distance l from the boundary Γ, the following estimate holds:

$$|u^d| \leqslant Ce^{-kcl}.$$

So, in Ω_2, u^d decreases exponentially with k; i.e., inside of Ω_2 the influence of u^- is not felt at all.

Further, from the above when $\lambda_1 < \lambda_2$, if the angle of incidence is so large that it satisfies (3.175), then there is total reflection at the boundary Γ, and no effect is transmitted into Ω_2.

3.6 The asymptotic solution of an elastic wave

For $\xi = (\xi_1, \xi_2, \xi_2) \in \mathbb{R}^3$ we set

$$(3.182) \qquad \boldsymbol{A}(\xi) = \begin{bmatrix} \alpha|\xi|^2 + \beta\xi_1^2 & \beta\xi_1\xi_2 & \beta\xi_1\xi_3 \\ \beta\xi_2\xi_1 & \alpha|\xi|^2 + \beta\xi_2^2 & \beta\xi_2\xi_3 \\ \beta\xi_3\xi_1 & \beta\xi_3\xi_2 & \alpha|\xi|^2 + \beta\xi_3^2 \end{bmatrix}.$$

· Further, we can express the elastic wave (1.31) by

$$(3.183) \qquad \rho\frac{\partial^2 \boldsymbol{u}}{\partial t^2} = \boldsymbol{A}\left(\frac{\partial}{\partial x}\right)\boldsymbol{u} + \boldsymbol{f}.$$

When $\xi \neq 0$, the eigenvalues of $\boldsymbol{A}(\xi)$ are $(\alpha + \beta)|\xi|^2$ and $\alpha|\xi|^2$, with the eigenspace, $\boldsymbol{E}_1(\xi)$, corresponding to $(\alpha + \beta)|\xi|^2$ given by

$$(3.184) \qquad \boldsymbol{E}_1(\xi) = \{p\xi; p \in \mathbb{C}\}$$

and the eigenspace, $\boldsymbol{E}_2(\xi)$, corresponding to $\alpha|\xi|^2$ given by

$$(3.185) \qquad \boldsymbol{E}_2(\xi) = \{\eta \in \mathbb{C}^3; (\eta, \xi) = 0\},$$

where $(\cdot, \cdot)$ denotes the inner product in $\mathbb{C}^3$.

Now set

$$(3.186) \qquad \lambda_1 = \sqrt{\frac{\alpha + \beta}{\rho}}, \quad \lambda_2 = \sqrt{\frac{\alpha}{\rho}}.$$

Next, we assume that the real-valued smooth function $\varphi(x)$ satisfies

$$|\nabla\varphi(x)|^2 = 1.$$

What we need to find is a function

(3.187) $$\boldsymbol{u}(t, x; k) = e^{ik(\lambda t - \varphi(x))}\boldsymbol{v}(t, x; k)$$

with

(3.188) $$\boldsymbol{v}(t, x; k) = \sum_{l=0}^{N}(ik)^{-l}\boldsymbol{v}_l(t, x),$$

such that as $k \to \infty$ asymptotically it satisfies

$$\boldsymbol{L}\boldsymbol{u} = \rho\frac{\partial^2}{\partial t^2}\boldsymbol{u} - \boldsymbol{A}\left(\frac{\partial}{\partial x}\right)\boldsymbol{u} = 0,$$

where $\boldsymbol{v}_l$ is a function that takes its value in $\mathbb{C}^3$.

If we now differentiate the $\boldsymbol{u}$ given by (3.187), then

(3.189) $$e^{-ik(\lambda t - \varphi(x))}\boldsymbol{L}\boldsymbol{u} = (ik)^2[\rho\lambda^2 I - \boldsymbol{A}(\nabla\varphi)]\boldsymbol{v} + (ik)\boldsymbol{T}\boldsymbol{v} + \boldsymbol{L}\boldsymbol{v},$$

where $\boldsymbol{T}$ is a first order partial differential operator given by

(3.190) $$\boldsymbol{T}\boldsymbol{v} = T\boldsymbol{v} + \beta(\text{div}\,\boldsymbol{v})\text{grad}\,\varphi + \beta\,\text{grad}\,(\text{grad}\,\varphi\cdot\boldsymbol{v}).$$

Here, we have set $T = 2\rho\lambda\dfrac{\partial}{\partial t} + 2\alpha\nabla\varphi\cdot\nabla + \alpha\Delta\varphi$ with $\nabla\varphi\cdot\nabla = \displaystyle\sum_{j=1}^{3}\frac{\partial\varphi}{\partial x_j}\frac{\partial}{\partial x_j}$.

So, let us substitute (3.188) for the $\boldsymbol{v}$ in the right-hand side of (3.189). Then, by rearranging and grouping the powers of ik, we see that

$$\begin{aligned}
\text{the r.h.s of (3.189)} = {}& (ik)^2[\rho\lambda^2 I - \boldsymbol{A}(\nabla\varphi)]\boldsymbol{v}_0 \\
&+ ik\{\boldsymbol{T}\boldsymbol{v}_0 + (\rho\lambda^2 I - \boldsymbol{A}(\nabla\varphi))\boldsymbol{v}_1\} \\
&+ (ik)^0\{\boldsymbol{L}\boldsymbol{v}_0 + \boldsymbol{T}\boldsymbol{v}_1 + (\rho\lambda^2 I - \boldsymbol{A}(\nabla\varphi))\boldsymbol{v}_2\} \\
&+ (ik)^{-1}\{\boldsymbol{L}\boldsymbol{v}_1 + \boldsymbol{T}\boldsymbol{v}_2 + (\rho\lambda^2 I - \boldsymbol{A}(\nabla\varphi))\boldsymbol{v}_3\} + \cdots \\
&+ (ik)^{-l}\{\boldsymbol{L}\boldsymbol{v}_l + \boldsymbol{T}\boldsymbol{v}_{l+1} + (\rho\lambda^2 I - \boldsymbol{A}(\nabla\varphi))\boldsymbol{v}_{l+2}\} + \cdots \\
&+ (ik)^{-N+1}\{\boldsymbol{L}\boldsymbol{v}_{N-1} + \boldsymbol{T}\boldsymbol{v}_N\} + (ik)^{-N}\boldsymbol{L}\boldsymbol{v}_N.
\end{aligned}$$

We want to select the $\boldsymbol{v}_j$ in such a way that the above coefficients become zero in order. Therefore, by considering the coefficient of $(ik)^2$ it is desired that

$$(3.191) \qquad (\rho\lambda^2 I - \boldsymbol{A}(\nabla\varphi))\boldsymbol{v}_0 = 0.$$

However, for $\boldsymbol{v}_0$ to be non-zero, we must have that $\rho\lambda^2$ is an eigenvalue of $\boldsymbol{A}(\nabla\varphi)$. But $|\nabla\varphi(x)|^2 = 1$, so either $\rho\lambda^2 = \alpha + \beta$ or $\rho\lambda^2 = \alpha$; that is, λ is equal to either λ_1 or λ_2.

First consider the case of $\lambda = \lambda_1$. From (3.184), we have that $\boldsymbol{v}_0(t, x) \in \boldsymbol{E}_1(\nabla\varphi(x))$ and so we can write

$$(3.192) \qquad \begin{cases} \boldsymbol{v}_0(t, x) = p_0(t, x)(\nabla\varphi)(x), \\ p_0(t, x) \text{ is a complex-valued function.} \end{cases}$$

In what follows, let

$$\boldsymbol{g}(x) = \nabla\varphi(x),$$

and choose smooth $\mathbb{R}^3$-valued functions $\boldsymbol{h}(x), \boldsymbol{j}(x) \in \boldsymbol{E}_2(\nabla\varphi(x))$ such that for each x the system $\{\boldsymbol{g}(x), \boldsymbol{h}(x), \boldsymbol{j}(x)\}$ becomes an orthonormal basis for $\mathbb{R}^3$. For $\boldsymbol{v}_0$ of the form in (3.192) a direct calculation shows that

$$(3.193) \qquad \boldsymbol{T}\boldsymbol{v}_0 = (Tp_0 + \beta\nabla\varphi \cdot \nabla p_0 + \beta\Delta\varphi p_0)\boldsymbol{g} + p_0 T\boldsymbol{g} + \beta\nabla p_0.$$

The $\boldsymbol{g}(x)$-coordinate of $\boldsymbol{T}\boldsymbol{v}_0$ is

$$Tp_0 + \beta\nabla\varphi \cdot \nabla p_0 + \beta\Delta\varphi p_0 + \beta\nabla\varphi \cdot \nabla p_0 + p_0 T\boldsymbol{g} \cdot \boldsymbol{g}.$$

Now, we should point out that $T\boldsymbol{g}\cdot\boldsymbol{g} = 2\alpha \sum_{j,h=1}^{3} \dfrac{\partial^2\varphi}{\partial x_j \partial x_h} \dfrac{\partial\varphi}{\partial x_j} \dfrac{\partial\varphi}{\partial x_h} + \alpha\Delta\varphi$. Therefore, if $p_0(t, x)$ satisfies

$$(3.194) \qquad T_1 p_0 = 2\rho\frac{\partial p_0}{\partial t} + 2(\alpha + \beta)\nabla\varphi \cdot \nabla p_0 + a_{1,\varphi}(x)p_0 = 0,$$

then

$$(3.195) \qquad (\boldsymbol{T}\boldsymbol{v}_0)(t, x) \in \boldsymbol{E}_2((\nabla\varphi)(x)),$$

where

$$a_{1,\varphi} = (2\alpha + \beta)\Delta\varphi + 2\alpha \sum_{j,h=1}^{3} \frac{\partial^2\varphi}{\partial x_j \partial x_h} \frac{\partial\varphi}{\partial x_j} \frac{\partial\varphi}{\partial x_h}.$$

Hence, we can determine $\boldsymbol{v}_0$ by (3.192) with p_0 satisfying (3.194).

Next, we want to find v_1 in the form

$$v_1(t, x) = p_1(t, x)g(x) + q_1(t, x)h(x) + r_1(t, x)j(x).$$

If p_0 satisfies (3.192), since (3.195) holds, q_1 and r_1 are uniquely determined by

(3.196) $$(\rho\lambda_1^2 I - A(\nabla\varphi))v_1 = -Tv_0.$$

So, if q_1 and r_1 are chosen in this way, then regardless of how we choose p_1, we have that the coefficient of ik in the right-hand side of (3.189) is zero.

From our consideration of v_0, since the g-coordinate $T(p_1 g)$ is $T_1 p_1$, if we can determine $p_1(t, x)$ in such a way that

(3.197) $$T_1 p_1 = -Lv_0 \cdot g - T(q_1 h + r_1 j) \cdot g$$

is satisfied, then the coefficient of $(ik)^0$ belongs to $E_2(\nabla\varphi(x))$ for all x.

Next, we would like to determine v_2 of the form

$$v_2(t, x) = p_2(t, x)g(x) + q_2(t, x)h(x) + r_2(t, x)j(x).$$

First, we choose q_2 and r_2 in such a way that the following is satisfied:

$$Lv_0 + Tv_1 + (\rho\lambda_1^2 I - A(\nabla\varphi))v_2 = 0.$$

Next, we need to determine p_2 so that the following holds:

$$T_1 p_2 = -Lv_1 \cdot g - T(q_2 h + r_2 j) \cdot g.$$

By repeating the above procedure, we can show in order that the coefficients in the right-hand side of (3.189) are zero, and so obtain that

$$Lu = e^{ik(\lambda_1 t - \varphi(x))}\{(ik)^{-N+1}(Lv_{N-1} + Tv_N) + (ik)^{-N}Lv_N\}.$$

Now, if k is sufficiently large, then Lu is reasonably close to zero. Also, the central part of u is the first term, and this is

(3.198) $$e^{ik(\lambda_1 t - \varphi(x))}p_0(t, x)\nabla\varphi(x).$$

As in the consideration of the asymptotic solution of the wave equation, (3.198) is a wave that progresses with velocity λ_1 in the direction $\nabla\varphi$. Also, from of our understanding of the amplitude function, the oscillation at each point is the same as the progression direction of the wave. That is, the wave is a longitudinal wave, a P-wave.

Next, we consider the case $\lambda = \lambda_2$. Since we must have in this case that

(3.199) $$\boldsymbol{v}_0(t,x) \in \boldsymbol{E}_2(\nabla\varphi(x)),$$

we set

(3.200) $$\boldsymbol{v}_0(t,x) = q_0(t,x)\boldsymbol{h}(x) + r_0(t,x)\boldsymbol{j}(x).$$

For this form of $\boldsymbol{v}_0$, we may again apply (3.190). If we consider the $E_2(\nabla\varphi)$ component of $\boldsymbol{T}\boldsymbol{v}_0$, then, since $\beta(\text{div}\,\boldsymbol{v}_0)\nabla\varphi \in E_1(\nabla\varphi)$ and $\nabla\varphi \cdot \boldsymbol{v}_0 = 0$, the part of $\boldsymbol{T}\boldsymbol{v}_0$ that is orthogonal to $\boldsymbol{g}$ is

$$(Tq_0 + q_0 T\boldsymbol{h}\cdot\boldsymbol{h} + r_0 T\boldsymbol{j}\cdot\boldsymbol{h})\boldsymbol{h} + (Tr_0 + r_0 T\boldsymbol{j}\cdot\boldsymbol{j} + q_0 T\boldsymbol{h}\cdot\boldsymbol{j})\boldsymbol{j}.$$

Therefore, it is sufficient that q_0 and r_0 satisfy the equations

(3.201) $$\begin{cases} Tq_0 + a_{11}q_0 + a_{12}r_0 = 0, \\ Tr_0 + a_{21}q_0 + a_{22}r_0 = 0, \end{cases}$$

where

$$a_{11} = T\boldsymbol{h}\cdot\boldsymbol{h}, \quad a_{12} = T\boldsymbol{j}\cdot\boldsymbol{h}, \quad a_{21} = T\boldsymbol{h}\cdot\boldsymbol{j}, \quad a_{22} = T\boldsymbol{j}\cdot\boldsymbol{j},$$

and these functions can be determined from φ.

Next, to determine $\boldsymbol{v}_1 = p_1\boldsymbol{g} + q_1\boldsymbol{h} + r_1\boldsymbol{j}$, we first set

$$p_1 = -(\beta\text{div}\,\boldsymbol{v}_0 + q_0 T\boldsymbol{h}\cdot\boldsymbol{g} + r_0 T\boldsymbol{j}\cdot\boldsymbol{g}).$$

Then it is sufficient to take q_1 and r_1 as the solutions of the simultaneous equations

(3.202) $$\begin{cases} Tq_1 + a_{11}q_1 + a_{12}r_1 = -\boldsymbol{L}\boldsymbol{v}_0\cdot\boldsymbol{h}, \\ Tr_1 + a_{21}q_1 + a_{22}r_1 = -\boldsymbol{L}\boldsymbol{v}_0\cdot\boldsymbol{j}. \end{cases}$$

Continuing in the above manner, we can determine in order $\boldsymbol{v}_l = p_l\boldsymbol{g} + q_l\boldsymbol{h} + r_l\boldsymbol{j}$ $(l = 2, 3, \ldots)$.

In this way, the asymptotic solution can be constructed with the principal part given by

$$e^{ik(\lambda_2 t - \varphi(x))}\{q_0(t,x)\boldsymbol{h}(x) + r_0(t,x)\boldsymbol{j}(x)\}.$$

Now, the wave expressed by the principal part is a wave that moves with speed λ_2 and the oscillation at each point is perpendicular to the direction of progress, $\nabla\varphi$, of the wave. In other words, this wave is a transverse wave; i.e., an S-wave.

In conclusion, an elastic equation has two waves with two different speeds, called an S-wave and a P-wave.

Chapter summary.

3.1 When the parameter k is large, using the phase function and the amplitude function we can construct concretely a solution that is suitably close to the true solution.

3.2 The phase function satisfies the *Eikonal* equation.

3.3 The amplitude function satisfies the transport equation.

3.4 The direction of transmission of a singularity is determined by the *Eikonal* equation.

3.5 Using the asymptotic solution for the wave equation, mathematically, we can show the observed properties of geometric optics.

3.6 The P-wave and S-wave that accompany an earthquake can be understood precisely by using the techniques of the asymptotic solution.

Exercises

1. Let $a_{11} \geqslant a_{22} \geqslant a_{33} > 0$ be constants and define the operator P by
$$P = \frac{\partial^2}{\partial t^2} - a_{11}\frac{\partial^2}{\partial x_1^2} - a_{22}\frac{\partial^2}{\partial x_2^2} - a_{33}\frac{\partial^2}{\partial x_3^2}.$$
(1) Find the asymptotic solution of the form of (3.2) and (3.3) for $P[u] = 0$.
(2) $\{\xi \in \mathbb{R}^3; a_{11}\xi_1^2 + a_{22}\xi_2^2 + a_{33}\xi_3^2 = 1\}$ is termed the *slowness surface* of the operator P; explain the rationale behind this statement.

2. Consider the Maxwell equations (1.20)–(1.23) for $q = 0$ and $\boldsymbol{j} = 0$. For (1.30) and div $\boldsymbol{A} = 0$, construct the asymptotic solution of the form
$$\boldsymbol{A}(t, x; k) = e^{ik(t-\psi(x))} \sum_{j=0}^{N} \boldsymbol{v}_j(t, x)(ik)^{-j}.$$

In addition, if we assume that $(\boldsymbol{A}, 0)$ is the asymptotic Lorentz gauge, find the asymptotic solution of the Maxwell equations with (1.24) and (1.25). With this asymptotic solution investigate the properties of electromagnetic waves.

3. Consider the elastic equation in a domain Ω with boundary, and suppose that the Dirichlet boundary condition holds; i.e.,
$$\boldsymbol{u}|_{\partial\Omega} = 0.$$

When the wave hits the boundary the asymptotic solution is of the form (3.187) and (3.188); construct the asymptotic solution corresponding to the reflected wave.

4. Consider the asymptotic solution $u(t, x; k)$ of the form (3.96) corresponding to the wave equation and with $\varphi(x) = -|x|$. Further, suppose that at $t = 0$ the support of u is $\{x; 1 \leqslant |x| \leqslant 2\}$. Then, for $t > 0$, in what range of t is $u(t, x; k)$ valid as an asymptotic solution?

CHAPTER 4

Local Energy of the Wave Equation

In Chapter 3, from the local point of view, we investigated the behaviour of the solution of the wave equation with respect to both time and space. In this chapter, we study from the global perspective the behaviour of the solution on the exterior of a bounded object, again with respect to time and space. In fact, with time, all of the oscillations are transmitted into the distance.

4.1 Local energy

We begin by assuming that $\mathcal{O} \subset \mathbb{R}^3$ is a bounded, open set with a smooth boundary Γ. Further, we assume that

$$(4.1) \qquad \Omega = \mathbb{R}^3 - \overline{\mathcal{O}}$$

is connected. We mean by this condition that all the external points of $\mathcal{O}$ are strung out towards infinity.

We consider the following initial boundary value problem of the wave equation:

$$(4.2) \qquad \begin{cases} \Box u(t, x) = 0 \quad \text{in } (0, \infty) \times \Omega, \\ u(t, x) = 0 \quad \text{on } (0, \infty) \times \Gamma, \\ u(0, x) = f_0(x), \quad \dfrac{\partial u}{\partial t}(0, x) = f_1(x). \end{cases}$$

The above problem is set up so that the wave is not transmitted into the interior of $\mathcal{O}$; that is, during the propagation of the wave, $\mathcal{O}$ exists as an "obstacle" to this propagation.

The solution u of (4.2) conserves energy in the way discussed in §2.2(e). To be more precise, suppose we define the total energy at time t of u to be

$$E(u; t) = \frac{1}{2} \int_\Omega \{|\nabla u(t, x)|^2 + |u_t(t, x)|^2\} dx.$$

Then the following holds:

$$E(u;t) = E(u;0) = \frac{1}{2} \int_{\Omega} \{|\nabla f_0(x)|^2 + |f_1(x)|^2\} dx \quad \forall t \geqslant 0.$$

Clearly, energy conservation holds in the case of $\Omega = \mathbb{R}^3$.

So, when $\Omega = \mathbb{R}^3$, we consider the case for which the initial data $f = \{f_0, f_1\}$ with $\rho > 0$ satisfies

$$\operatorname{supp} f = \bigcup_{j=0}^{1} \operatorname{supp} f_j \subset \{|x| \leqslant \rho\}.$$

Then, from the Huygens principle, we have that

$$u(t, x) = 0 \quad \text{for } |x| \leqslant t - \rho.$$

So, from the above, for each fixed $R > 0$, in the region $|x| \leqslant R$ the energy is zero for $t \geqslant R + \rho$. Further, expression (1.41) shows that the solution moves in a straight manner into the distance with velocity 1.

So now, the natural question to ask is, "Does this also hold for the obstacle case?" Due to the arguments of §3.5(h), the wave is reflected on hitting the obstacle. As in a bay, the wave will have repeated reflections, which might not be able to escape out of the bay. So, this phenomenon is not that easy to understand.

Further, although it was not explained in Chapter 3, the wave has what is termed diffraction. That is, the wave bends into the shadowy area, as understood from the perspective of geometric optics. As the frequency increases, the study of geometric optics gives a very good approximation to the behaviour of the wave, but even if the quantity is small, diffraction will still occur. Therefore, due to this diffraction, the solution also propagates behind the obstacle. So, this type of solution encompasses rather complicated behaviour.

However, for Ω that satisfies condition (4.1) with the initial data that has a finite energy, the effect with time gradually drifts into the distance, while the effect that remains in the bounded region becomes small. In order to demonstrate these facts we need to introduce the following concepts.

DEFINITION 4.1. Suppose that u is a solution of (4.2). For $R > 0$, let us set $\Omega_R = \Omega \cap \{x; |x| < R\}$ and define $E(t; u, R)$ by

$$E(t; u, R) = \frac{1}{2} \int_{\Omega_R} \{|\nabla u(t, x)|^2 + |u_t(t, x)|^2\} dx.$$

We call $E(t; u, R)$ the *local energy* of the solution in Ω_R at time t. For future reference, set

$$E(t; u, \infty) = \frac{1}{2} \int_\Omega \{|\nabla u(t, x)|^2 + |u_t(t, x)|^2\} dx.$$

In this book we shall not give a proof, but just write down the facts discussed above in the next theorem.

THEOREM 4.2. *Suppose that* $f = \{f_1, f_2\}$ *is a finite energy data. Then for arbitrary* $R > 0$ *the solution* u *of* (4.2) *with* f *satisfies*

$$\lim_{t \to \infty} E(t; u, R) = 0.$$

As a reference to the proof of the theorem we refer the reader to Lax and Phillips [6, Chapter 5, Theorem 2.1]. In fact, in Chapter 5, Section 6 [*ibid*], the history of the research behind this theorem is discussed. From the theorem, for all the initial data with finite energy, as $t \to \infty$ the local energy in Ω_R of the solution of (4.2) decays to zero. Now, as the local energy decays to zero as $t \to \infty$, what exactly is the speed of this decay? To be able to consider this problem we need to introduce the space of initial data, namely,

$$\mathcal{H}_\rho(\Omega) = \{\{u_0, u_1\}; u_0 \in H^2(\Omega) \cap H_0^1(\Omega), u_1 \in H_0^1(\Omega), \text{supp } u_j \subset \overline{\Omega}_\rho\}.$$

Now, suppose that $\overline{\mathcal{O}} \subset \{x; |x| < \rho_0\}$. So, for a fixed $R > \rho_0$, when $\{u_0, u_1\} \in \mathcal{H}_R(\Omega)$ how large must t be so that $E(t; u, R) \leqslant \frac{1}{2} E(0; u, R)$? This is an important question if we wish to consider things from the point of view of the decay of the local energy.

As previously noted, the propagation of the solution with a high frequency can be approximated by geometric optics. Therefore, the geometric optic light ray is caught in $\mathcal{O}$ and if it has difficulty escaping, the wave expressed by the solution of (4.2) can also be thought of as having difficulty escaping. So, we introduce a geometric property of $\mathcal{O}$ via the next definition.

DEFINITION 4.3. $\mathcal{O}$ *is said to be* non-trapping *if for an arbitrary* $R > \rho_0$ *there exists a positive constant* T_R *such that no matter from which point* x *of* Ω_R *and in what direction* $\xi \in S^2 = \{\xi \in \mathbb{R}^3; |\xi| = 1\}$ *emerges, a geometric optic light ray within time* T_R *will always leave* Ω_R.

$\mathcal{O}$ *is said to be* trapping *when it is not non-trapping.*

For an arbitrary $T > 0$, from this definition, if $\mathcal{O}$ is trapping, there always exists a geometric optic light ray that starts from some point of Ω_R and after time T still remains in Ω_R.

First we give an illustrative example of a non-trapping $\mathcal{O}$.

EXAMPLE 4.4. (Convex set)

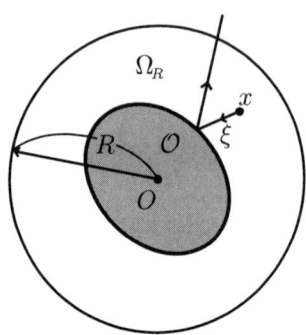

FIGURE 4.1

EXAMPLE 4.5. (Star-shaped set) A set $\mathcal{O}$ is said to be *star-shaped,* if there exists a point p of $\mathcal{O}$ with the following property: for an arbitrary $x \in \Gamma$, $\overrightarrow{px} \cdot \nu(x) \geqslant 0$ is true, where $\nu(x)$ is an outward unit normal vector of Γ at $x \in \Gamma$ with respect to $\mathcal{O}$. As we shall soon see, a star-shaped set is non-trapping.

Suppose that a light ray emerges from x and, as in Figure 4.2, is reflected at x^1, x^2. Now, we take q to be a point on the reflected light ray from x^2. In addition, let q' lie on the half-line that extends the segment x^1x^2 from x^2 and satisfies

$$|x^2q| = |x^2q'| \ (= l).$$

Next, set $\overrightarrow{x^1x^2}/|x^1x^2| = \xi$ and $\overrightarrow{x^2q}/|x^2q| = \eta$. From the law of reflection in geometric optics, we have that $\eta = \xi - 2(\xi \cdot \nu)\nu$ and $(\xi, \nu) \leqslant 0$.

Therefore, since

$$|pq'|^2 = |\overrightarrow{px^2} + l\xi|^2 = |\overrightarrow{px^2}|^2 + 2l\overrightarrow{px^2} \cdot \xi + l^2,$$
$$|pq|^2 = |\overrightarrow{px^2} + l\eta|^2 = |\overrightarrow{px^2}|^2 + 2l\overrightarrow{px^2} \cdot \eta + l^2,$$

we obtain that

$$|pq|^2 - |pq'|^2 = 2l\,\overrightarrow{px^2}\cdot(-2(\xi\cdot\nu)\nu) = -4l(\xi\cdot\nu)(\overrightarrow{px^2}\cdot\nu) \geqslant 0,$$

and so $|pq| \geqslant |pq'|$.

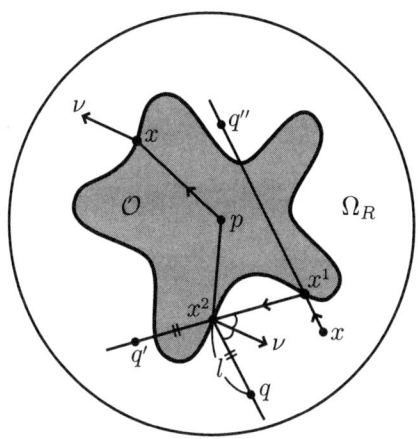

FIGURE 4.2

In a similar way, if we take q'' to lie on the elongation of the segment xx^1 from x^1 with the provision that $|x^1 q''| = |x^1 q'|$, then $|pq''| \leqslant |pq'|$. So, the length of the light ray measured from x up to q is equal to $|xq''|$.

Now, suppose that q lies in the interior of Ω_R; then $|pq| \leqslant 2R$, and hence $|pq''| \leqslant 2R$. Therefore, $|xq''| \leqslant 4R$; that is, after time $4R$, q cannot remain in Ω_R.

Even for a large number of reflected points, the above arguments are still valid, and hence $\mathcal{O}$ is non-trapping.

Having looked at an example of non-trapping, let us now give an example of a trapping set.

EXAMPLE 4.6. (The case where $\mathcal{O}$ is made up of $\mathcal{O}_1$ and $\mathcal{O}_2$ which have no point in common.) If we take $a_j \in \Gamma = \partial\mathcal{O}_j$ $(j = 1, 2)$ so that the distance between $\mathcal{O}_1$ and $\mathcal{O}_2$ is dis$(\mathcal{O}_1, \mathcal{O}_2) = |a_1 a_2|$, then $a_1 a_2$ is perpendicular to Γ_j at a_j. Hence, the light ray that emerges from a_1 in the direction of a_2 is reflected at a_2 and returns to a_1, then again moves in the direction of a_2. Therefore, since we always bounce back and forth between a_1 and a_2, $\mathcal{O}$ is trapping.

FIGURE 4.3

In Example 4.6, it is the case that we move back and forth between a_1 and a_2 along the geometric optic path. Now, by using the methods of the asymptotic solution discussed in Chapter 3, we can create an asymptotic solution $u(t, x; k)$ of (4.2) that has support only in a small neighbourhood of the segment $a_1 a_2$. If, in advance, we fix a $T > 0$, then this solution is such that it exists in the region $t \in [0, T]$. For high frequencies, since $\Box u$, itself, gradually becomes small, in the time interval $[0, T]$, the asymptotic solution approaches the real solution. So, for Example 4.6 we can say the following: "For a fixed $T > 0$, choose an arbitrary $\varepsilon > 0$. Then, for the initial value $f = \{f_0, f_1\} \in \mathcal{H}_R(\Omega)$, we have that $\int_\Omega \{|\nabla f_0|^2 + |f_1|^2\}dx > 0$ is satisfied, and, moreover, the solution u of (4.2), with the initial value f, satisfies

(4.3) $E(t; u, R) \geqslant (1 - \varepsilon)E(0; u, \infty), \ \forall t \in [0, T]$."

For any trapping obstacle, there always exists an initial value $f \in \mathcal{H}_R(\Omega)$ that is non-zero and satisfies (4.3) for an arbitrary $T > 0$. This fact shows that if $\mathcal{O}$ is trapping, regardless of the size of $T > 0$, then a solution of (4.2) always exists and the amount of energy seeping from Ω_R over a period of only T is less than ε.

Now, in order to determine how long it takes for more than half the original energy to leak out of Ω_R, we are very much dependent on the properties of the initial data. So, this problem needs to be examined on a case by case basis, and some cases become quite involved. Therefore, in the next section we will look at the following problem: "For the initial data in $\mathcal{H}_R(\Omega)$, what types of Ω are there so that, after a fixed period of time, some fixed ratio of energy always dissipates from Ω_R?"

4.2 A uniform decay of the local energy

To address the problem stated above, we begin by giving the following definition.

DEFINITION 4.7. The initial boundary value problem (4.2) has a uniform decay of local energy if, for arbitrary $R > \rho_0$, we can find a positive function $p_R(t)$, defined on $[0, \infty)$, that satisfies

$$(4.4) \qquad \lim_{t \to \infty} p_R(t) = 0,$$

and for arbitrary initial data $\{u_0, u_1\} \in \mathcal{H}_R(\Omega)$, an estimate of the local energy of the solution u is

$$(4.5) \qquad E(t; u, R) \leqslant p_R(t) E(0; u, \infty) \qquad (t \geqslant 0).$$

As was explained in the previous section, for a uniform decay of the local energy to occur $\mathcal{O}$ must be non-trapping. Conversely, if $\mathcal{O}$ is non-trapping, it is known that the local energy decays uniformly. However, a proof of the latter statement is quite difficult, and so we will restrict ourselves here to showing that the local energy decays uniformly in the relatively simple case (within all non-trapping obstacles) of an obstacle formed by a star-shaped domain. The proof we give is due to Morawetz, [4].

THEOREM 4.8. *Suppose that $\mathcal{O}$ is a bounded star-shaped domain with boundary Γ, and fix an $R > \rho_0$. Then, when the initial data $\{f_0, f_1\}$ belongs to $\mathcal{H}_R(\Omega)$, the solution u at time $t > R$ has local energy that satisfies the following estimate:*

$$(4.6) \qquad E(t; u, R) \leqslant \frac{C(R)}{t - R} E(0; u, \infty),$$

where $C(R)$ is a constant that depends on R but is independent of u.

PROOF. Without lost of generality, we assume that $\mathcal{O}$ is star-shaped with respect to the origin. Further, we assume the solution of (4.2) is real-valued. Our intention is to integrate

$$U = \left(\sum_{j=1}^{3} x_j u_{x_j} + t u_t + u \right) \left(u_{tt} - \sum_{l=1}^{3} u_{x_l x_l} \right)$$

over $(0, T) \times \Omega$.

To this end, we rearrange the right-hand side of the above, on a term by term basis as follows:

$$-x_j u_{x_j} u_{x_j x_j} = -\left(\frac{1}{2}x_j u_{x_j}^2\right)_{x_j} + \frac{1}{2}u_{x_j}^2 \quad (j=1,2,3),$$

$$-x_j u_{x_j} u_{x_l x_l} = -(x_j u_{x_j} u_{x_l})_{x_l} + \left(\frac{1}{2}x_j u_{x_l}^2\right)_{x_j} - \frac{1}{2}u_{x_l}^2 \quad (j \neq l),$$

$$-t u_t u_{x_j x_j} = -(t u_t u_{x_j})_{x_j} + \left(\frac{1}{2}t u_{x_j}^2\right)_t - \frac{1}{2}u_{x_j}^2,$$

$$x_j u_{x_j} u_{tt} = (x_j u_{x_j} u_t)_t - \left(\frac{1}{2}x_j u_t^2\right)_{x_j} + \frac{1}{2}u_t^2,$$

$$t u_t u_{tt} = \frac{1}{2}(t u_t^2)_t - \frac{1}{2}u_t^2,$$

$$-u u_{x_j x_j} = -(u u_{x_j})_{x_j} + u_{x_j}^2 \quad (j=1,2,3),$$

$$u u_{tt} = (u u_t)_t - u_t^2.$$

From this, we can rewrite U as

$$U = (X_1)_{x_1} + (X_2)_{x_2} + (X_3)_{x_3} + (X_0)_t$$

where

$$X_1 = \frac{1}{2}x_1(-u_{x_1}^2 + u_{x_2}^2 + u_{x_3}^2 - u_t^2) - x_2 u_{x_1} u_{x_2} - x_3 u_{x_1} u_{x_3}$$
$$- t u_t u_{x_1} - u u_{x_1},$$

$$X_2 = \frac{1}{2}x_2(-u_{x_2}^2 + u_{x_3}^2 + u_{x_1}^2 - u_t^2) - x_3 u_{x_2} u_{x_3} - x_1 u_{x_2} u_{x_1}$$
$$- t u_t u_{x_2} - u u_{x_2},$$

$$X_3 = \frac{1}{2}x_3(-u_{x_3}^2 + u_{x_1}^2 + u_{x_2}^2 - u_t^2) - x_1 u_{x_3} u_{x_1} - x_2 u_{x_3} u_{x_2}$$
$$- t u_t u_{x_3} - u u_{x_3},$$

$$X_0 = \frac{1}{2}t(u_t^2 + u_{x_1}^2 + u_{x_2}^2 + u_{x_3}^2) + x_1 u_t u_{x_1} + x_2 u_t u_{x_2} + x_3 u_t u_{x_3}$$
$$+ u u_t.$$

Now, if we assume that

$$(4.7) \qquad \mathrm{supp}\, f_0,\ \mathrm{supp}\, f_1 \subset \{x \in \overline{\Omega}; |x| \leqslant R\},$$

then since the speed of propagation of the solution is 1, for $t > 0$ we have

(4.8) $\operatorname{supp} u(t, \cdot) \subset \{x \in \overline{\Omega}; |x| \leqslant R + t\} = \Omega_{R+t}.$

Next, we integrate U over $(0, t) \times \Omega$ and apply the divergence theorem. From (4.8), for $j = 1, 2, 3$ the following holds:

$$\int_{(0,t) \times \Omega} (X_j)_{x_j}\, dt dx = -\int_0^t dt \int_\Gamma X_j \nu_j\, dS,$$

where $\nu = (\nu_1, \nu_2, \nu_3)$ denotes the outward unit normal vector with respect to $\mathcal{O}$ of Γ.

Also, we obtain that

$$\int_{(0,t) \times \Omega} (X_0)_t\, dt dx = \int_\Omega X_0(t, x)\, dx - \int_\Omega X_0(0, x)\, dx.$$

Therefore, we see that

(4.9)

$$0 = \int_\Omega X_0(t, x)\, dx - \int_\Omega X_0(0, x)\, dx$$
$$- \int_0^t dt \int_\Gamma \left(\sum_{j=1}^3 X_j \nu_j \right) dS.$$

Now, since $u = 0$ on Γ, $(u_{x_1}, u_{x_2}, u_{x_3})$ on Γ is also a normal vector of Γ, namely, if we set $\sum_{j=1}^3 \nu_j \dfrac{\partial u}{\partial x_j} = \dfrac{\partial u}{\partial \nu}$, then on Γ we have that

(4.10) $\dfrac{\partial u}{\partial x_j} = \nu_j \dfrac{\partial u}{\partial \nu} \qquad (j = 1, 2, 3).$

On the other hand, since $u = 0$ on Γ, we also have $u_t = 0$ on Γ. Substituting this and then calculating $\sum\limits_{j=1}^{3} X_j \nu_j$, on Γ we get that

$$
\begin{aligned}
\sum_{j=1}^{3} X_j \nu_j &= \left(\frac{\partial u}{\partial \nu}\right)^2 \left[\left\{ \frac{1}{2}x_1(-\nu_1^2 + \nu_2^2 + \nu_3^2) - x_2\nu_1\nu_2 - x_3\nu_1\nu_3 \right\} \nu_1 \right. \\
&\quad + \left\{ \frac{1}{2}x_2(-\nu_2^2 + \nu_3^2 + \nu_1^2) - x_3\nu_2\nu_3 - x_1\nu_2\nu_1 \right\} \nu_2 \\
&\quad + \left. \left\{ \frac{1}{2}x_3(-\nu_3^2 + \nu_1^2 + \nu_2^2) - x_1\nu_3\nu_1 - x_2\nu_3\nu_2 \right\} \nu_3 \right] \\
&= \left(\frac{\partial u}{\partial \nu}\right)^2 \left[\frac{1}{2}x_1\nu_1(-\nu_1^2 - \nu_2^2 - \nu_3^2) \right. \\
&\quad + \frac{1}{2}x_2\nu_2(-\nu_1^2 - \nu_2^2 - \nu_3^2) \\
&\quad + \left. \frac{1}{2}x_3\nu_3(-\nu_1^2 - \nu_2^2 - \nu_3^2) \right] \\
&= -\frac{1}{2}\left(\frac{\partial u}{\partial \nu}\right)^2 (\nu_1^2 + \nu_2^2 + \nu_3^2)(x_1\nu_1 + x_2\nu_2 + x_3\nu_3).
\end{aligned}
$$

Since $\mathcal{O}$ is a star-shaped domain with respect to the origin, we see from the definition that $x_1\nu_1 + x_2\nu_2 + x_3\nu_3 \geqslant 0$.

Hence, from the above, on Γ we have that

$$
\sum_{j=1}^{3} X_j \nu_j \leqslant 0.
$$

Using this estimate in (4.9), we obtain that

$$
(4.11) \qquad \int_\Omega X_0(t, x)dx \leqslant \int_\Omega X_0(0, x)dx.
$$

Let us now estimate the right-hand side of (4.11). Noting that $\{f_0, f_1\} \in \mathcal{H}_R(\Omega)$, we see that

$$
\int_\Omega \left| \sum_{j=1}^{3} x_j u_t u_{x_j} \right| dx \leqslant \frac{R}{2} \int_\Omega \{|u_t(0, x)|^2 + |\nabla u(0, x)|^2\}dx
$$

$$
= \frac{R}{2}(\|f_1\|_{L^2(\Omega)}^2 + \|\nabla f_0\|_{L^2(\Omega)}^2).
$$

By using inequality (4.18) for $u(0, \cdot)$ (this inequality is part of Lemma 4.9 which we shall prove later), we obtain that

$$\int_{\Omega} |u(0, x)u_t(0, x)|dx \leqslant \frac{C_R}{2}\{||\nabla f_0||^2_{L^2(\Omega)} + ||f_1||^2_{L^2(\Omega)}\}.$$

From the above, we can choose a $C_1(R)$ that is determined by $\mathcal{O}$ and R, and hence we get that

$$(4.12) \qquad \int_{\Omega} X_0(0, x)dx \leqslant C_1(R)\{||\nabla f_0||^2_{L^2(\Omega)} + ||f_1||^2_{L^2(\Omega)}\}$$

$$= C_1(R)E(0; u, \infty).$$

To simplify matters subsequently, let us denote $E(0; u, \infty)$ by E_0.

Next, let us estimate the left-hand side of (4.11). By considering the domain of influence of the initial value, we obtain that

$$(4.13) \qquad \operatorname{supp} u(t, \cdot) \subset \Omega_{R+t}.$$

Now, the following estimate holds:

$$\sum_{j=1}^{3} x_j u_t u_{x_j} = |x|u_t \left(\frac{x}{|x|} \cdot \nabla u\right) \leqslant \frac{|x|}{2}\{|u_t|^2 + |\nabla u|^2\}.$$

Hence, the integral over Ω_R of $X_0 - uu_t$ can be estimated as

$$(4.14) \qquad \int_{\Omega_R} \left\{\frac{t}{2}\left(u_t^2 + \sum_{j=1}^{3} u_{x_j}^2\right) + \sum_{j=1}^{3} x_j u_t u_{x_j}\right\} dx$$

$$\geqslant \frac{t-R}{2} \int_{\Omega_R} (u_t^2 + |\nabla u|^2)dx;$$

while for the integral over $\Omega - \Omega_R$, by taking account of (4.8) we see that

$$(4.15)$$

$$\int_{|x|>R} \left\{\frac{t}{2}\left(u_t^2 + \sum_{j=1}^{3} u_{x_j}^2\right) + \sum_{j=1}^{3} x_j u_t u_{x_j}\right\} dx$$

$$\geqslant \frac{t-(t+R)}{2} \int_{|x|\geqslant R} (u_t^2 + |\nabla u|^2)dx = -R\,E(t; u, \infty) = -RE_0.$$

Now, from Lemma 4.10, which we will prove shortly, when $f \in \mathcal{H}_R(\Omega)$ the following holds:

$$||u(t, \cdot)||^2_{L^2(\Omega)} \leqslant C_2(R)E_0,$$

where $C_2(R)$ is a constant that is determined from $\mathcal{O}$ and R.

Using this, we get that

(4.16)
$$\left| \int_\Omega u u_t dx \right| \leqslant \frac{1}{2} \{ \|u(t,\cdot)\|_{L^2(\Omega)}^2 + \|u_t(t,\cdot)\|_{L^2(\Omega)}^2 \} \leqslant (C_2(R)+1)E_0.$$

From (4.14), (4.15) and (4.16), we obtain that

(4.17)
$$\int_\Omega X_0(t,x)dx \geqslant \frac{t-R}{2} \int_{\Omega_R} \{ |u_t(t,x)|^2 + |\nabla u(t,x)|^2 \} dx$$
$$- (R + C_2(R) + 1)E_0.$$

Due to the above, if we substitute (4.12) and (4.17) into both sides, respectively, of (4.11), we have that

$$(t-R)E(t;u,R) - (R + C_2(R) + 1)E_0 \leqslant C_1(R)E_0.$$

So, if we set $C(R) = 2C_1(R) + (R + C_2(R) + 1)$, we see that

$$E(t;u,R) \leqslant \frac{C(R)}{t-R}E_0.$$

This final inequality is exactly what was required.

$\square$

To complete this section we state and prove the two lemmas we alluded to and used in the preceding proof.

LEMMA 4.9. *Suppose that $\mathcal{O}$ is star-shaped with respect to the origin. Further, suppose that $w(x) \in H_0^1(\Omega)$. If we assume that $R > \rho_0$, then the following estimate holds:*

(4.18)
$$\|w\|_{L^2(\Omega_R)} \leqslant \frac{R^{3/2}}{\sqrt{3r_0}} \|\nabla w\|_{L^2(\Omega)},$$

where $r_0 = \text{dis}\,(0, \Gamma)$.

PROOF. Suppose that $w(x) \in C^\infty(\overline{\Omega})$. For $\theta \in S^2$, we define an $r(\theta) > 0$ such that $r(\theta)\theta \in \Gamma$. For every $\theta \in S^2$, $r(\theta) \geqslant r_0$. Now, when $r > r(\theta)$, from

$$w(r\theta) = \int_{r(\theta)}^r \frac{\partial}{\partial s} w(s\theta) ds$$

and by using Schwarz's inequality, we see that

$$|w(r\theta)|^2 \leqslant \int_{r(\theta)}^r s^{-2} ds \int_{r(\theta)}^r s^2 |\nabla w(S\theta)|^2 \leqslant \frac{1}{r(\theta)} \int_{r(\theta)}^\infty |\nabla w(s\theta)|^2 s^2 ds.$$

Therefore, we have

$$
\begin{aligned}
||w||^2_{L^2(\Omega_R)} &= \int_{S^2} d\theta \int_{r(\theta)}^{R} |w(r\theta)|^2 r^2 dr \\
&\leqslant \int_{S^2} \left(\int_{r(\theta)}^{R} r^2 dr \right) \left(\frac{1}{r(\theta)} \int_{r(\theta)}^{\infty} |\nabla w(s\theta)|^2 s^2 ds \right) d\theta \\
&\leqslant \frac{R^3}{3} \int_{S^2} \frac{1}{r_0} \int_{r(\theta)}^{\infty} |\nabla w(s\theta)|^2 s^2 ds d\theta \leqslant \frac{R^3}{3r_0} ||\nabla w||^2_{L^2(\Omega)}.
\end{aligned}
$$

For $w \in H^1(\Omega)$, we take the sequence of elements in $C^\infty(\overline{\Omega})$ that converge to w in $H^1(\Omega)$, and then we apply the above result.

□

LEMMA 4.10. *For all $t > 0$, the solution u of (4.2) with the initial value $f \in \mathcal{H}_R(\Omega)$ satisfies*

(4.19) $$||u(t,\cdot)||^2_{L^2(\Omega)} \leqslant C_2(R)\{||\nabla f_0||^2_{L^2(\Omega)} + ||f_1||^2_{L^2(\Omega)}\},$$

where if we denote the coefficient of the right-hand side of (4.18) by C_R, then $C_2(R) = C_R + C_R^2$.

PROOF. Begin by fixing a $T > 0$, and then let $\Omega^{(0)} = \Omega_{R+T+1}$. By noting (4.13), u may be regarded as the solution of

(4.20) $$\begin{cases} \Box u = 0 & \text{in } (0,T) \times \Omega^{(0)}, \\ u = 0 & \text{on } (0,T) \times \partial\Omega^{(0)}, \\ u(0,x) = f_0(x), & u_t(0,x) = f_1(x). \end{cases}$$

Now suppose that $g \in H^3(\Omega^{(0)}) \cap H^1_0(\Omega^{(0)})$ is a solution of

(4.21) $$\begin{cases} \Delta g(x) = f_1(x) & \text{in } \Omega^{(0)}, \\ g = 0 & \text{on } \partial\Omega^{(0)}. \end{cases}$$

The existence of the solution g is guaranteed by Fredholm's alternating theorem (see Yosida [2], page 289).

In addition, we suppose that w is the solution of

(4.22) $$\begin{cases} \Box w = 0 & \text{in } (0,T) \times \Omega^{(0)}, \\ w = 0 & \text{on } (0,T) \times \partial\Omega^{(0)}, \\ w(0,x) = g(x), & w_t(0,x) = f_0(x). \end{cases}$$

Since in this problem the energy is conserved, we have that

$$
(4.23) \quad
\begin{aligned}
\int_{\Omega^{(0)}} &\{|\nabla w(t,x)|^2 + |w_t(t,x)|^2\}dx \\
&= \{||\nabla g(x)||^2_{L^2(\Omega^{(0)})} + ||f_0||^2_{L^2(\Omega^{(0)})}\}.
\end{aligned}
$$

By partially differentiating (4.22) with respect to t, w_t satisfies

$$
(4.24) \quad
\begin{cases}
\Box w_t = 0 & \text{in } (0,T) \times \Omega^{(0)}, \\
w_t = 0 & \text{on } (0,T) \times \partial\Omega^{(0)}, \\
w_t(0,x) = f_0(x), \quad (w_t)_t(0,x) = \dfrac{\partial^2 w}{\partial t^2}(0,x).
\end{cases}
$$

Since $\Box w = 0$, we see that $w_{tt}(0,x) = \Delta w(0,x) = \Delta g(x) = f_1(x)$. Therefore, (4.20) and (4.24) are exactly the same initial boundary value problem. So gathering these facts together and using the uniqueness of the solution, we have that

$$
u(t,x) = w_t(t,x) \qquad ((t,x) \in (0,T) \times \Omega^{(0)}).
$$

Hence, by considering (4.13) and using (4.23), we see that

$$
(4.25) \quad
\begin{aligned}
||u(t,\cdot)||^2_{L^2(\Omega)} = \int_{\Omega^{(0)}} |u(t,x)|^2 dx &= \int_{\Omega^{(0)}} |w_t(t,x)|^2 dx \\
&\leqslant ||\nabla g||^2_{L^2(\Omega^{(0)})} + ||f_0(x)||^2_{L^2(\Omega^{(0)})}.
\end{aligned}
$$

Next, we estimate the right-hand side of the above. Using (4.18), we have that

$$
(4.26) \quad ||f_0||^2_{L^2(\Omega^{(0)})} = ||f_0||^2_{L^2(\Omega_R)} \leqslant C_R ||\nabla f_0||^2_{L^2(\Omega)}.
$$

Also,

$$
\begin{aligned}
||\nabla g||^2_{L^2(\Omega^{(0)})} &= \int_{\Omega^{(0)}} \nabla g(x) \cdot \nabla g(x) dx \\
&= -\int_{\Omega^{(0)}} \Delta g(x) g(x) dx = -\int_{\Omega_R} f_1(x) g(x) dx.
\end{aligned}
$$

Now, by applying Schwarz's inequality, followed by the use of (4.18), we obtain that

$$
\begin{aligned}
\left|\int_{\Omega_R} f_1(x) g(x) dx\right| &\leqslant ||f_1||_{L^2(\Omega_R)} ||g||_{L^2(\Omega_R)} \\
&\leqslant ||f_1||_{L^2(\Omega)} C_R ||\nabla g||_{L^2(\Omega)}.
\end{aligned}
$$

From this, we get that

(4.27) $$\|\nabla g\|_{L^2(\Omega^{(0)})} \leqslant C_R^2 \|f_1\|_{L^2(\Omega_R)}.$$

Finally, by substituting (4.26) and (4.27) into the right-hand side of (4.25), we achieve our goal and obtain (4.19).

□

4.3 The exponential decay of the local energy

As noted in the previous section, for an arbitrary non-trapping obstacle it is known that the local energy decays uniformly.

When the dimension of the space is three, we noted in §1.3 that Huygens' principle holds. By using this fact, it is possible to see that when there is a uniform decay of the local energy, the speed of this decay is always exponential.

THEOREM 4.11. *For an obstacle $\mathcal{O}$, suppose the local energy decays uniformly. Then, there exist a positive constant α that is independent of R and a positive constant C that depends on R. With these constants, for arbitrary initial data $\{f_0, f_1\} \in \mathcal{H}_R(\Omega)$, the solution u has an estimate of local energy of the form,*

$$E(t; u, R) \leqslant Ce^{-\alpha t}E(0; u, R) \qquad (t \geqslant 0).$$

The proof of the above theorem will be divided into several steps. We begin with the next lemma, the proof of which we leave to the reader.

LEMMA 4.12. *Suppose that $f_0 \in H^2(\Omega) \cap H_0^1(\Omega)$ and $f_1 \in H^1(\Omega)$. Then, there exist $\widetilde{f_0} \in H^2(\mathbb{R}^3)$ and $\widetilde{f_1} \in H^1(\mathbb{R}^3)$ such that*

$$\widetilde{f_l}(x) = f_l(x) \quad a.e. \ x \in \Omega,$$

and for arbitrary $R \geqslant \rho_0$ the following hold:

$$\|\nabla \widetilde{f_0}\|_{L^2(B_R)} \leqslant C\|\nabla f_0\|_{L^2(\Omega_R)},$$
$$\|\widetilde{f_1}\|_{L^2(B_R)} \leqslant C\|f_1\|_{L^2(\Omega_R)},$$

where $B_R = \{x \in \mathbb{R}^3; |x| < R\}$ and C is a constant independent of f_0 and f_1.

Now we define a mapping $U(t)$ from $H_0^1(\Omega) \times L^2(\Omega)$ to itself by $U(t)f = \{u(t, \cdot), u_t(t, \cdot)\}$, where u is the solution of (4.2) with the initial data $f = \{f_0, f_1\}$. Further, suppose that $f = \{f_0, f_1\} \in$

$\mathcal{H}_R(\Omega)$, and finally that $T > 0$ is fixed. If we set $g = \{g_0, g_1\} = \{u(T, \cdot), u_t(T, \cdot)\}$, then by applying Lemma 4.12 we can find $\widetilde{g} = \{\widetilde{g}_0, \widetilde{g}_1\} \in H^2(\mathbb{R}^3) \times H^1(\mathbb{R}^3)$ such that for arbitrary $R > \rho_0$ the following holds:

(4.28)
$$\|\nabla \widetilde{g}_0\|_{L^2(B_R)} \leqslant C \|\nabla g_0\|_{L^2(\Omega_R)}, \qquad \|\widetilde{g}_1\|_{L^2(B_R)} \leqslant C \|g_1\|_{L^2(\Omega_R)}.$$

Now suppose that the initial surface is $t = T$, and $h(t, x)$ is the solution of the initial value problem with the initial value $\widetilde{g}$. That is to say, h is the solution of

(4.29)
$$\begin{cases} \Box h(t, x) = 0 & \text{in } (T, \infty) \times \mathbb{R}^3, \\ h(T, x) = \widetilde{g}_0(x), \quad h_t(T, x) = \widetilde{g}_1(x). \end{cases}$$

LEMMA 4.13. *If $T \geqslant R$, the solution h of (4.29) satisfies*

(4.30)
$$h(t, x) = 0 \qquad \text{for } |x| < t - (T + \rho_0).$$

PROOF. To begin with, let

$$k(t, x) = \begin{cases} u(t, x), & (t, x) \in (0, T] \times \Omega, \\ h(t, x), & (t, x) \in (T, \infty) \times \mathbb{R}^3. \end{cases}$$

From the method of choosing the initial condition of h, the following holds:

(4.31)
$$\Box k(t, x) = 0 \qquad \text{in } ((0, \infty) \times \mathbb{R}^3) \backslash ((0, T] \times \mathcal{O}).$$

For a fixed $\varepsilon > 0$, we take a $\chi_\varepsilon(t, x) \in C^\infty([0, \infty) \times \mathbb{R}^3)$ such that

$$\chi_\varepsilon(t, x) = \begin{cases} 1, & |x| \geqslant \rho_0 \text{ or } t \geqslant T + \varepsilon, \\ 0, & (t, x) \in [0, T] \times \mathcal{O}. \end{cases}$$

Moreover, suppose that at $t \in [0, T]$, we have $\chi_\varepsilon(t, x) = \chi_\varepsilon(0, x)$. Further, let $k = 0$ on $(0, T] \times \overline{\mathcal{O}}$ and suppose k is defined on $(0, \infty) \times \mathbb{R}^3$. So, we may set

(4.32)
$$\Box(\chi_\varepsilon k)(t, x) = r_\varepsilon(t, x).$$

Therefore, $\chi_\varepsilon k$ becomes the solution of the following initial problem:

$$\begin{cases} \Box v = r_\varepsilon & \text{in } (0, \infty) \times \mathbb{R}^3, \\ v(0, x) = \chi_\varepsilon(0, x) k(0, x), \quad v_t(0, x) = \chi_\varepsilon(0, x) k_t(0, x), \end{cases}$$

where v is an unknown function.

If $\chi_\varepsilon = 1$ on $((0, \infty) \times \mathbb{R}^3) \backslash ((0, T) \times \{x; |x| \leqslant \rho_0\})$, then by using (4.31) we see that

$$(4.33) \qquad \operatorname{supp} r_\varepsilon \subset [0, T + \varepsilon] \times \{x; |x| \leqslant \rho_0\}.$$

Also, from the fact that $f \in \mathcal{H}_R(\Omega)$, the following holds:

$$(4.34) \qquad \operatorname{supp} \chi_\varepsilon(0, \cdot) k(0, \cdot) \cup \operatorname{supp} \chi_\varepsilon(0, \cdot) k_t(0, \cdot) \subset \{x : |x| \leqslant R\},$$

where we have used the formulation of the solution of the wave equation that was given in §1.3.

Now, if we use (1.41) and (1.51), we obtain the following expression for $\chi_\varepsilon k$:

$$(4.35) \qquad \begin{aligned} \chi_\varepsilon k(t, x) = {} & \frac{\partial}{\partial t} M[\chi_\varepsilon k(0, \cdot)](t, x) + M[\chi_\varepsilon k_t(0, \cdot)](t, x) \\ & + \int_0^t M[r_\varepsilon(s, \cdot)](t - s, x) ds. \end{aligned}$$

From (4.34) and Huygens' principle, we get that

$$(4.36) \quad \frac{\partial}{\partial t} M[\chi_\varepsilon k(0, \cdot)](t, x) + M[\chi_\varepsilon k_t(0, \cdot)](t, x) = 0 \quad \text{for } |x| \leqslant t - R.$$

Also, from (4.33) the integral of the third term of the right-hand side of (4.35) is on the interval from zero to $T + \varepsilon$, so if $s \in [0, T + \varepsilon]$, then we have that

$$(4.37) \qquad M[r_\varepsilon(s, \cdot)](t - s, x) = 0 \quad \text{for } |x| \leqslant t - (T + \varepsilon) - \rho_0.$$

Therefore, by applying (4.36) and (4.37) to the expression of (4.35), for $T > R$ we see that

$$(4.38) \qquad \chi_\varepsilon(t, x) k(t, x) = 0 \quad \text{for } |x| \leqslant t - (T + \rho_0 + \varepsilon).$$

But, if we note that $\varepsilon > 0$ was chosen arbitrarily, (4.30) follows from (4.38).

$$\square$$

In $(T, \infty) \times \Omega$ we define the function w by

$$(4.39) \qquad w(t, x) = u(t, x) - h(t, x).$$

So, for w the following is true.

LEMMA 4.14. *The function $w(t, x)$ satisfies the following:*

$$(4.40) \qquad \operatorname{supp} w \subset \{(t, x); |x| \leqslant \rho_0 + t - T, \ t \geqslant T\},$$

$$(4.41) \qquad w(t, x) = 0 \quad on \ [T + 2\rho_0, \infty) \times \Gamma,$$

$$(4.42) \qquad E(t; w, \infty) \leqslant (4 + 2C)E(T; u, 5\rho_0),$$

where the C in (4.42) is the same as the one defined in Lemma 4.12.

PROOF. We have that $\Box w = 0$ holds on $\{(t, x); t \geqslant T, \ x \in \Omega\}$; on the other hand, $u(T, x) = h(T, x)$ and $u_t(T, x) = h_t(T, x)$ hold for $x \in \Omega$. Therefore, since $w(T, x) = w_t(T, x) = 0 \ (x \in \Omega)$, by considering the domain of dependence of the wave equation, then by using Theorem 2.4, and finally by noting that $\lambda_{\max} = 1$, we obtain the required (4.40).

Next, we prove (4.41). Since $\mathcal{O} \subset \{x; |x| \leqslant \rho_0\}$, then from (4.30) when $t \geqslant T + 2\rho_0$ we have $h(t, x) = 0$ for $x \in \overline{\mathcal{O}}$. Also, since u is the solution of (4.2), we get that $u|_\Gamma = 0$. Now, from the definition of w in (4.39), we obtain (4.41).

Finally, we prove (4.42). Since on $(T, \infty) \times \Omega$ we have $\Box w = 0$ and also that (4.41) holds, at $t \geqslant T + 2\rho_0$, w is the solution of

$$(4.43) \qquad \begin{cases} \Box w = 0 & \text{in } (T + 2\rho_0, \infty) \times \Omega, \\ w = 0 & \text{on } (T + 2\rho_0, \infty) \times \Gamma. \end{cases}$$

In fact, for (4.43) energy is conserved; i.e.,

$$(4.44) \qquad E(t; w, \infty) = E(T + 2\rho_0; w, \infty) \quad (t \geqslant T + 2\rho_0).$$

Now we estimate the right-hand side of (4.44). From (4.40),

$$(4.45) \qquad \operatorname{supp} w(T + \rho_0, \cdot) \cup \operatorname{supp} w_t(T + 2\rho_0, \cdot) \subset \Omega_{3\rho_0}.$$

If we replace $P = \Box$ in (2.30) and then note that $Z = 0$, we see that

$$(4.46) \qquad E(T + 2\rho_0; h, 3\rho_0) \leqslant E(T; h, 5\rho_0),$$

and from (4.29) we obtain that

$$(4.47) \qquad \begin{aligned} &\int_{|x| \leqslant 3\rho_0} \{|\nabla h(T + 2\rho_0, x)|^2 + |h_t(T + 2\rho_0, x)|^2\} dx \\ &\qquad \leqslant \int_{|x| \leqslant 5\rho_0} \{|\nabla \widetilde{g}_0(x)|^2 + |\widetilde{g}_1(x)|^2\} dx. \end{aligned}$$

Now,

$$E(T + 2\rho_0; w, 3\rho_0) \leqslant 2\{E(T + 2\rho_0; u, 3\rho_0) + E(T + 2\rho_0; h, 3\rho_0)\}.$$

Next, by estimating the term on the right-hand side of the above by (4.46) and (4.47), and finally by using (4.28), we see that (4.42) holds.

□

With the above preparations we show that the local energy decays exponentially. First, we assume that the uniform decay of the local energy is in the sense of Definition 4.7. Further, we take $T_1 = T + 2\rho_0$. From (4.30), if $T \geqslant R$, then from (4.41) we see that $w(T_1, \cdot)$ and $w_t(T_1, \cdot)$ are in $H_0^1(\Omega)$. Also, from $u(T_1, \cdot) \in H^2(\Omega)$ and $h(T_1, \cdot) \in H^2(\mathbb{R}^3)$, we have that $w(T_1, \cdot) \in H^2(\Omega)$. Further, from (4.40), we obtain that $\{w(T_1, \cdot), w_t(T_1, \cdot)\} \in \mathcal{H}_{3\rho_0}(\Omega)$.

Next, if we assume that $R \geqslant 5\rho_0$, and then substitute (4.5) into the right-hand side of (4.42), we get that

$$E(T_1; w, \infty) \leqslant (4 + 2C)p_R(T)E(0; u, \infty).$$

Collecting the above deliberations together yields the following proposition.

PROPOSITION 4.15. *Suppose that* $R \geqslant 5\rho_0$. *Choose* T *in such a way that* $T \geqslant R$ *and, moreover,* T *satisfies* $(4 + 2C)p_R(T) \leqslant \frac{1}{2}$. *Further, set* $T_1 = T + 2\rho_0$. *Then, for* $t \geqslant T_1$, *we have the decomposition*

$$u(t, x) = w(t, x) + h(t, x),$$

and h and w satisfy

$$h(t, x) = 0 \quad for \ |x| \leqslant \rho_0 + (t - T_1),$$
$$\{w(T_1, \cdot), w_t(T_1, \cdot)\} \in \mathcal{H}_{3\rho_0}(\Omega),$$
$$E(t; w, \infty) \leqslant \frac{1}{2}E(0; u, \infty).$$

From Proposition 4.15, if $t \geqslant T_1 + R$, then for $|x| \leqslant R$ we have that $h(t, x) = 0$. Therefore,

$$(4.48) \qquad E(t; u, R) = E(t; w, R) \leqslant E(T_1; w, \infty) \leqslant \frac{1}{2}E(0; u, \infty).$$

Next, we take the initial surface to be $t = T_1$. Then by considering the initial data to be $\{w(T_1, \cdot), w_t(T_1, \cdot)\} \in \mathcal{H}_{3\rho_0}(\Omega)$ and by applying the above result, at $t \geqslant T_1 + \rho$ we obtain that

$$E(T_1 + t; w, R) \leqslant \frac{1}{2} E(T_1; w, \infty) = \frac{1}{2} E(T_1; w, R) \leqslant \left(\frac{1}{2}\right)^2 E(0; u, \infty).$$

So, by using the first equality in (4.48), we see that

$$(4.49) \qquad E(t; u, R) \leqslant \left(\frac{1}{2}\right)^2 E(0; u, \infty) \qquad \forall t \geqslant 2T_1 + R.$$

The solution u satisfies the estimate above for arbitrary $f \in \mathcal{H}_R(\Omega)$. Hence, with (4.49) and the initial surface set to be $t = T_1$, and then by applying the initial data $\{w(T_1, \cdot), w_t(T_1, \cdot)\} \in \mathcal{H}_{3\rho_0}(\Omega)$, we have from the above estimate that

$$E(T_1 + t; w, R) \leqslant \left(\frac{1}{2}\right)^2 E(T_1; w, \infty), \qquad \forall t \geqslant 2T_1 + R.$$

Again using (4.48), we see that

$$E(t; u, R) \leqslant \left(\frac{1}{2}\right)^3 E(0; u, \infty), \qquad \forall t \geqslant 3T_1 + R.$$

By repeating this procedure, we eventually obtain

$$(4.50) \qquad E(t; u, R) \leqslant \left(\frac{1}{2}\right)^m E(0; u, \infty), \qquad \forall t \geqslant mT_1 + R.$$

Therefore, if we assume $\alpha > 0$ is such that $e^{-\alpha T_1} = \frac{1}{2}$, then, by suitably choosing $C > 0$, we can express (4.50) as

$$E(t; u, R) \geqslant Ce^{-\alpha t} E(0; u, \infty), \qquad \forall t \geqslant 0.$$

Since the above is the desired estimate, we have proven Theorem 4.11.

$\square$

Chapter summary.

4.1   The local energy decays to zero.

4.2   At the exterior of a non-trapping obstacle, the local energy decays uniformly.

4.3   If there is a uniform decay, then the speed of the decay is exponential.

Exercises

1. (Penrose's mushroom) In the figure below, suppose that the curve ABC is the upper half of an ellipse with foci F and F'. Show that the geometric optic light ray that emerges from either $\mathcal{R}$ or $\mathcal{R}'$ cannot escape from the figure.

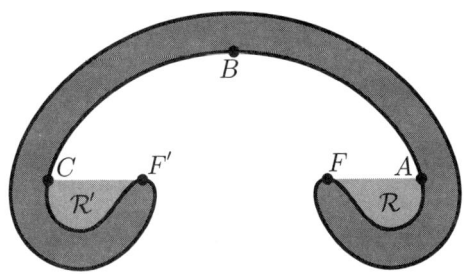

PENROSE'S MUSHROOM

2. Suppose that $h_0(x) \geqslant \alpha > 0$, and consider the initial value problem of the wave equation with friction given by

$$Pu = \frac{\partial u^2}{\partial t^2} - \Delta u + h_0(x)\frac{\partial u}{\partial t} = 0.$$

Show that the total energy decays in the following way:

$$E(t; u, \infty) \leqslant C(1+t)^{-1} E(0; u, \infty).$$

(Hint: Use $(Pu)\bar{u}_t$, $(Pu)(t\bar{u}_t + \lambda\bar{u})$ $(\lambda > 0)$.)

Perspectives on Current
Research in Mathematics

In July of 1990, I had the good fortune to attend an international research conference entitled *25 Years of Microlocal Analysis*, which was held very close to the borders of Austria and Switzerland. At this conference, Prof. L. Gårding, one of the preeminent researchers in the theory of partial differential equations, delivered a lecture on the topic of (the) *History of the Mathematics of Double Refraction*. The phenomenon of "double refraction" can be typically observed by looking at what happens to a letter once it passes through a piece of calcite to its underside. As the name suggests, the letter that appears on the underside is seen in double, as if the observer has a case of double vision. In terms of a ray of light entering this piece of calcite, this phenomena is the result of this light ray being divided into several rays.

Prof. Gårding's lecture concerned itself with the history of the mathematical research behind double refraction, starting with Huygens' "Traité de la lumière", published in 1690, and continuing to the current day. The lecture resonated deeply with me, and so I asked for a hard copy of the lecture. The paper of the lecture was subsequently published in *Archive for History of Exact Sciences*, **40** (1989), 355–385.

To be honest, I am quite unfamiliar with the history of the sciences or even with mathematics. Having said that, I was quite struck and surprised by the contents of the lecture. Professor Gårding in his history of research into double refraction noted that contributions have been made by: Huygens, Fresnel, Hamilton, Lamé, Sonya Kovalevskaya, Volterra, Grünwald, Fredholm, Zeilen, Herglotz, Petrovsky; moreover, there also exists collaborative work on this theme by Atiyah, Bott and Gårding.

As a matter of interest, Huygens was born 10 years before New-
ton. If we take a close look at the above list of mathematicians, who
lived before and after the discovery of differential and integral calcu-
lus, then what is interesting to see is not that so many great mathematicians carried out some research into light, but rather that the
extraordinary phenomena contingent on light continually attracted a
considerable number of mathematicians. Further, in their endeavours
to try and explain from a mathematical point of view these phenomena, they inexorably created new segments of mathematics.

In fact, Professor Gårding's survey showed that the progress of
mathematical research into double refraction stimulates progress in
analysis, in particular, advances in the theory of partial differential
equations, and *vice versa*. Now, the study of light precedes the theory
of partial differential equations, so it is to be expected that this would
incessantly give rise to new questions in the theory of partial differential equations, which in turn lead to new insights and developments.
It is my opinion that this will continue for some time to come. To
all intents and purposes, we know that the relationship between the
study of wave phenomena, of which light is an exemplary representative, and the theory of partial differential equations is difficult to
sever.

Continuing the above line of thought, we very briefly mention a
few areas of research that are presently evolving within mathematics
and which are based on problems discussed in this book.

1. Microlocal analysis.

On examining the propagation of a wave, if the wave has a singularity, then this above all else is the most interesting thing for us to
consider. Therefore, we must first see how this singularity is transmitted. As we discussed in Chapter 3, §3.4(b), if the singularity is on
some smooth surface, then with time this singularity will be propagated in a normal direction to the surface.

In the investigation of the singularity, it is important to know
not only the position of the singularity but also the direction of the
singularity. For a distribution on $\mathbb{R}^n$, we consider the set of the position x of the singularity and the direction ξ of the singularity at x,
so we can write $(x, \xi) \in T^*(\mathbb{R}^n)$. Then the analysis of $(x, \xi) \in T^*(\mathbb{R}^n)$
is a pivotal part of the study of the partial differential equations.
The analysis that concerns itself with such $(x, \xi) \in T^*(\mathbb{R}^n)$ is called
microlocal analysis.

If we consider partial differential operators in a microlocal sense, it is possible to obtain a more precise analysis. Currently, microlocal analysis is used in all aspects of analysis.

2. Fourier integral operator.

By using the Fourier transform, a function, defined on $\mathbb{R}^n$ and with certain properties, can be expressed as

$$f(x) = \left(\frac{1}{2\pi}\right)^{n/2} \int_{\mathbb{R}^n} e^{ix\xi} \widehat{f}(\xi)d\xi.$$

So, by means of the above expression, the linear problem basically reduces to the case with data $e^{ix\xi}$.

For example, we look at the initial value problem for a hyperbolic operator of second order, P. Given $u(0,x) = 0$ and $u_t(0,x) = f(x)$, if we can construct a G such that

$$PG(t,x,\xi) = 0, \quad G(0,x,\xi) = 0, \quad G_t(0,x,\xi) = e^{ix\xi},$$

then the desired solution u can be written (at least formally) as

$$u(t,x) = \left(\frac{1}{2\pi}\right)^{n/2} \int_{\mathbb{R}^n} G(t,x,\xi)\widehat{f}(\xi)d\xi$$

$$= \left(\frac{1}{2\pi}\right)^{n} \int_{\mathbb{R}^n \times \mathbb{R}^n} G(t,x,\xi)e^{-iy\xi}f(y)dyd\xi.$$

If we now use the methods for asymptotic solutions discussed in Chapter 3, then by making $|\xi|$ sufficiently large, we can form an approximate solution of G to a good degree of precision. Since the approximate solution of G is explicitly formed, the approximate solution, in turn, becomes useful in the study of the exact properties of the solution u. Also, in the study of the singularity of u it suffices to examine only the component for which ξ is large.

The generalization of the operator formed in the above fashion is the *Fourier integral operator*. To be precise, it is the operator A defined by

$$Af(x) = \int_{\mathbb{R}^n \times \mathbb{R}^N} \exp(i\varphi(x,\theta,y))a(x,\theta,y)f(y)dyd\theta,$$

where θ is a variable in $\mathbb{R}^N$. In particular, if $\varphi(x,\theta,y) = (x-y)\cdot\theta$ ($\theta \in \mathbb{R}^n$), then the operator is called the *pseudo-differential operator*.

The methods that use Fourier integral operators are extremely strong. An example of this is the tremendous progress that has been made in the study of the asymptotic distribution (Weyl formula) of the eigenvalues of an elliptic operator in a bounded domain.

A particular enhancement of the Fourier integral operator is the FBI operator. In fact, by means of the FBI operator it is possible to derive an accurate analysis of the diffraction of a wave. Until the appearance of the FBI operator, the mathematical study of diffraction phenomena was quite limited to the extent that the most intrinsic parts could not be resolved. However, the results of Lebeau, which made use of the FBI operator, showed that for light that impacts on a convex object, understood in a narrow sense, the intensity of the light in the part that is under shadow is

$$\exp\left(-k^{1/3}\int_0^l \alpha(s)ds\right) \times \text{ the intensity of the original source,}$$

where $\alpha(s)$ is a positive-valued function that is determined from the curvature of the object, l denotes the length of geodesic from the point at which light comes into contact with the object and up to the observed point, and k is the frequency. In the cases of high frequencies, the extent to which a very small amount of light is transmitted to the part in shadow has been examined in detail. Such a precise analysis is made possible by the use of the FBI operator.

3. Scattering theory.

In Chapter 4, we briefly discussed the concept of a wave hitting some object and thereupon being variously dispersed. By investigating the nature of this scattering, we can gain insight into the object that causes the scattering. Not surprisingly, *scattering theory* is the name given to the theory that deals with the relationship between the dispersion of a wave and the circumstances that are the primary factors of this scattering. An example of an application of scattering theory is the investigation of the way a wave is transmitted within the earth. The subsequent analysis allows us to better understand the inner parts of the earth.

As the above example illustrates, such questions are very interesting problems not only as research topics in mathematics, where the relevant branch of mathematics is vibrant and expanding, but they are also important problems from the standpoint of practical applications in engineering. It is my intention to study scattering theory with respect to wave equations in my subsequent book, *Scattering theory.*

Bibliography

Cited references

[1] V.I. Smirnov, *A course of higher mathematics,* vol. II, Chapter VII, Section 17 (1964), Pergamon Press, Oxford, England (translated by D.E. Brown).

[2] K. Yosida, *Functional analysis* (1980), Springer-Verlag, Berlin, New York (sixth edition).

[3] J.L. Lions and E. Magenes, *Problèmes aux limites non homogènes et applications,* vol. I (1968), Dunford, Paris.
(In Chapter 2, we need to make use of some results on elliptic equations, these can be found in the above book.)

[4] C.S. Morawetz, *The decay of solutions of the exterior intial-boundary value problem for the wave equation,* Comm. Pure Appl. Math **14** (1961), 561–568.

[5] C.S. Morawetz, *Exponential decay of solutions of the wave equation,* Comm. Pure Appl. Math **19** (1966), 439–444.
(Theorem 4.8 and Lemma 4.10 are the substance of the above two papers.)

[6] P.D. Lax and R.S. Philips, *Scattering theory* (1989), Academic Press, New York (revised edition).
(Theorem 4.2 is proved by Theorem 2.1 of Chapter 5 of the above book. As mentioned in the text of the present book, the decay of local energy is extremely important in scattering theory; in the section *Notes and Remarks*, at the end of Chapter 5 of the above book, a brief history of the research in this area is given.)

General references

[1] I.G. Petrovsky, *Lectures on partial differential equations* (1950), Interscience, New York.
(The above book is an introductory text on partial differential equations based on lectures given at Moscow State University by one of the foremost Russian authorities on partial differential equations. Hyperbolic equations are the central focus of the book.)

[2] R. Courant and D. Hilbert, *Methods of Mathematical Physics,* vol. 2 (1962), Interscience, New York.
(This book covers the general theory of partial differential equations in relation to their importance to many concepts in physics. Courant wrote this treatise based on deliberations with his mentor Hilbert; it imparts the sense of research in partial differential equations.)

[3]   R. Sakamoto, *Hyperbolic boundary value problems* (1982), Cambridge University Press, Cambridge, New York.
(In this book, also on the general theory of partial differential equations, methods of dealing with hyperbolic equations, based on the named author's work, are developed.)

[4]   S. Mizohata, *The theory of partial differential equations* (1973), Cambridge University Press, Cambridge, New York.
(In the above, initial boundary value problems are considered for general hyperbolic equations.)

[5]   J. Rauch, *Partial differential equations* (1991), Springer-Verlag, New York, Tokyo.
(This book deals with the overall theory of partial differential equations.)

[6]   L. Hörmander, *The analysis of linear partial operators,* vols. I, II, III, IV (1985), Springer-Verlag, Berlin, Heidelberg, New York.
(The most current methods and results are presented in these four volumes that cover consummately the theory of partial differential equations.)

Solutions to the Exercises

Chapter 1.

1. (1) We begin with the change of variables $\xi = x - t$ and $\eta = x + t$, and so $x = (\xi + \eta)/2$ and $t = (\eta - \xi)/2$. Then, if we set $\widetilde{u}(\xi, \eta) = u((\xi + \eta)/2, (\eta - \xi)/2)$, we see that

$$\frac{\partial^2 \widetilde{u}}{\partial \xi \partial \eta}(\xi, \eta) = \frac{1}{4}(\Box u)((\xi + \eta)/2, (\eta - \xi)/2) = 0.$$

Hence, we can write $\widetilde{u}(\xi, \eta) = f(\xi) + g(\eta)$.

(2) It suffices to choose f and g so that the initial conditions are satisfied. So, since

$$f(x) + g(x) = \varphi(x), \quad -f'(x) + g'(x) = \psi(x),$$

we obtain that $f'(x) = (\varphi'(x) - \psi(x))/2$ and that $g'(x) = (\varphi'(x) + \psi(x))/2$.

Now, integrate and adjust the constant accordingly.

(3) If we assume $u(t, 0) = f(-t) + g(t) = 0$, then we must have that $g(t) = -f(-t)$.

(4) Let f be as in (3). The requirement is that $f(x) - f(-x) = \varphi(x)$ $(x > 0)$ and $f'(x) - f'(-x) = \psi(x)$ $(x > 0)$. Now, if we assume that f is defined on all of $\mathbb{R}$, then $f(x) - f(-x)$ is an odd function. So, if we denote the extensions of the odd functions φ and ψ to all of $\mathbb{R}$ by $\widetilde{\varphi}(x)$ and $\widetilde{\psi}(x)$, respectively, we obtain that

$$f(l) = \frac{1}{2} \left\{ \widetilde{\varphi}(l) + \int_0^l \widetilde{\psi}(s) \, ds \right\} \quad (l \in \mathbb{R}).$$

Hence, we have that

$$u(t, x) = \frac{1}{2} \left\{ \widetilde{\varphi}(x + t) + \widetilde{\varphi}(x - t) + \int_{x-t}^{x+t} \widetilde{\psi}(s) \, ds \right\}.$$

In the case when $x \geqslant t$, in the above equations, $\widetilde{\varphi}$ and $\widetilde{\psi}$ are equal to φ and ψ, respectively. While, when $0 < x < t$,

$$u(t,x) = \frac{1}{2}\left\{\varphi(x+t) - \varphi(t-x) + \int_{t-x}^{t+x} \psi(s)\,ds\right\}.$$

2. Assume that L is the length of the string and ρ is the density of the string. At the point x, with this distance being measured from the ceiling, the force that is exerted by the string from this point and up to the bottom end is given by $g\rho(L-x)$, where g is the gravitational constant. Therefore, the tension at this point is $g\rho(L-x)$. Then, with deference to the methods in §1.1(a), we obtain that

$$\rho\frac{\partial^2 u}{\partial t^2} - \frac{\partial}{\partial x}\left(g\rho(L-x)\frac{\partial u}{\partial x}\right) = 0.$$

3. This is just a direct calculation.

4. (1) For $Pu(t,x) = 0$, if we set $\widetilde{u}(t,x) = u(t,\sqrt{a}\,x)$, then we get that $\Box\widetilde{u}(t,x) = 0$, where $\sqrt{a}\,x = (\sqrt{a_{11}}\,x_1, \sqrt{a_{22}}\,x_2, \sqrt{a_{33}}\,x_3)$. Therefore, if we take $\widetilde{u}$ to be the solution of

$$\Box\widetilde{u} = 0, \quad \widetilde{u}(0,x) = 0, \quad \widetilde{u}_t(0,x) = \varphi(\sqrt{a}\,x),$$

then we can define u in terms of the above relations. Now, we express $\widetilde{u}$ by means of (1.41); then by a change of variable we obtain the following expression for u:

$$u(t,x) = \frac{1}{4\pi t\sqrt{|a|}}\int_{d_A(x,\xi)=t} \varphi(\xi)\,dS_\xi,$$

where $|a| = a_{11}a_{22}a_{33}$,

$$d_A(x,\xi) = \left\{\frac{(x_1-\xi_1)^2}{a_{11}} + \frac{(x_2-\xi_2)^2}{a_{22}} + \frac{(x_3-\xi_3)^2}{a_{33}}\right\}^{1/2},$$

and dS_ξ denotes the area element of the surface $d_A(x,\xi) = t$.

(2) We have that

$$\operatorname{supp} u(t,\cdot) \subset \{x;\ \text{for some } |y| \leqslant \varepsilon, \quad d_A(x,y) = t\}.$$

But, this is contained in $\{x;\ \sqrt{a_{33}}\,t - \varepsilon \leqslant |x| \leqslant \sqrt{a_{11}}\,t + \varepsilon\}$.

5. Set $u = \operatorname{div} \boldsymbol{B}$. Then, since $\Box u = -\operatorname{div}\operatorname{rot}\boldsymbol{j} = 0$ and, moreover, $u(0,x) = u_t(0,x) = 0$, we obtain that $u \equiv 0$. Now, (1.23)

follows from the definition of $\boldsymbol{E}$. Further, because div $\boldsymbol{B} = 0$, it follows that rot rot $\boldsymbol{B} = -\Delta\boldsymbol{B}$, and so we have that

$$
\begin{aligned}
\text{rot } \boldsymbol{E} &= \int_0^t (-\Delta\boldsymbol{B} + \text{rot } \boldsymbol{j})ds + \text{rot } \boldsymbol{E}_0 \\
&= \int_0^t \left(-\frac{\partial^2 \boldsymbol{B}}{\partial t^2} \right) ds - \boldsymbol{B}_t(0,x) = -\frac{\partial \boldsymbol{B}}{\partial t}(t,x).
\end{aligned}
$$

So, $\partial_l(\text{div } \boldsymbol{E}) = \text{div } \boldsymbol{j} = \partial_t q$. Then, since

$$
q(0,x) = \text{div } \boldsymbol{E}(0,x),
$$

we see that div $\boldsymbol{E} = q$.

Chapter 2.

1. Define $\widetilde{u}$ by $u(t,x) = u(t' + \Phi(x'), x') = \widetilde{u}(t', x')$. From the equality $u(t,x) = \widetilde{u}(t - \Phi(x), x)$, we have that

$$
\frac{\partial u}{\partial t} = \frac{\partial \widetilde{u}}{\partial t'}, \quad \frac{\partial u}{\partial x_j} = \frac{\partial \widetilde{u}}{\partial t'} \left(-\frac{\partial \Phi}{\partial x_j} \right) + \frac{\partial \widetilde{u}}{\partial x_j}.
$$

Therefore,

$$
\begin{aligned}
P[u] = & \left\{ 1 - 2\sum_{j=1}^n h_j \frac{\partial \Phi}{\partial x_j} - \sum_{j,l=1}^n a_{jl}(t,x) \left(-\frac{\partial \Phi}{\partial x_j} \right) \left(-\frac{\partial \Phi}{\partial x_l} \right) \right\} \frac{\partial^2 \widetilde{u}}{\partial t'^2} \\
& + 2\sum_{j=1}^n \left\{ h_j - \sum_{l=1}^n a_{jl} \left(-\frac{\partial \Phi}{\partial x_l} \right) \right\} \frac{\partial^2 \widetilde{u}}{\partial t' \partial x_j'} - \sum_{j,l=1}^n a_{jl} \frac{\partial^2 \widetilde{u}}{\partial x_j' \partial x_l'} \\
& + \text{ an order operator of first order.}
\end{aligned}
$$

So, the characteristic equation is

$$
\begin{aligned}
\widetilde{p}_0(t',x',\lambda',\xi') &= p_0\left(t',x',1,-\frac{\partial \Phi}{\partial x} \right) \lambda'^2 \\
& \quad + 2\sum_{j=1}^n \left\{ h_j - \sum_{l=1}^n a_{jl} \left(-\frac{\partial \Phi}{\partial x_l} \right) \right\} \lambda' \xi_l' - \sum a_{jl} \xi_j' \xi_l' \\
&= p_0\left(t',x',\lambda',\xi' - \lambda'\frac{\partial \Phi}{\partial x} \right).
\end{aligned}
$$

The roots of the quadratic equation with respect to λ' are the characteristic roots.

In the region for which $\Phi(x)$ satisfies

$$p_0(t', x', 1, -\Phi_x(x')) > 0,$$

the transformed operator $\widetilde{P}$ is hyperbolic.

2. Take the inner product of both sides with respect to $\overline{\boldsymbol{u}}_t$ and then integrate over $(0, t) \times \mathbb{R}^3$. So, we have that

$$\mathrm{Re}\left\{\alpha\Delta\boldsymbol{u}\cdot\overline{\boldsymbol{u}}_t + \beta\,\mathrm{grad}\,\mathrm{div}\,\boldsymbol{u}\cdot\overline{\boldsymbol{u}}_t\right\}$$

$$= -\frac{1}{2}\frac{\partial}{\partial t}\left\{\alpha\sum_{j=1}^{3}|\nabla u_j|^2\right\} + \mathrm{Re}\sum_{l=1}^{3}\frac{\partial}{\partial x_l}\left\{\sum_{j=1}^{3}\frac{\partial u_j}{\partial x_l}\frac{\partial\overline{u}_j}{\partial t}\right\}$$

$$- \frac{1}{2}\frac{\partial}{\partial t}\left\{\beta\left|\sum_{j=1}^{3}\frac{\partial u_j}{\partial x_j}\right|^2\right\} + \mathrm{Re}\frac{\partial}{\partial x_l}\left\{\beta\sum_{j=1}^{3}\frac{\partial u_j}{\partial x_j}\frac{\partial u_l}{\partial t}\right\}.$$

Now, if we set

$$E(\boldsymbol{u}; t) = \frac{1}{2}\left(\alpha\sum_{j=1}^{3}|\nabla u_j|^2 + \beta\left|\sum_{j=1}^{3}\frac{\partial u_j}{\partial x_j}\right|^2\right),$$

then it follows that

$$E(\boldsymbol{u}; t) \leqslant e^t\left\{E(\boldsymbol{u}; 0) + \int_0^t \|f(s, \cdot)\|_{L^2(\mathbb{R}^3)}^2\,ds\right\}.$$

With regard to the existence of the solution, for $\lambda > 0$, if we prove the existence of the solution of

$(*)$ $\alpha\Delta\boldsymbol{u} + \beta\,\mathrm{grad}\,\mathrm{div}\,\boldsymbol{u} - \lambda\boldsymbol{u} = \boldsymbol{g},$

then we can apply the Hille-Yosida theorem. The method of proof of the existence of the solution of $(*)$ is essentially the same as that for Δ with a single unknown function.

3. First,

$$\int_{\mathbb{R}^n} A_j(x)\frac{\partial\boldsymbol{u}}{\partial x_j}\cdot\overline{\boldsymbol{u}}\,dx = -\int_{\mathbb{R}^n}\boldsymbol{u}\cdot\left(\frac{\partial}{\partial x_j}{}^t A_j(x)\overline{\boldsymbol{u}}\right)dx$$

$$= -\int_{\mathbb{R}^n}\boldsymbol{u}\cdot\overline{{}^t A_j(x)\frac{\partial\boldsymbol{u}}{\partial x_j}}\,dx$$

$$-\int_{\mathbb{R}^n}\boldsymbol{u}\cdot({}^t A_j)_{x_j}\overline{\boldsymbol{u}}\,dx.$$

Now, since $\overline{{}^t A_j} = A_j$, we see that

$$2\mathrm{Re} \int_{\mathbb{R}^n} A_j \frac{\partial u}{\partial x_j} \cdot \overline{u}\, dx = -\int_{\mathbb{R}^n} u \cdot ({}^t A_j)_{x_j} \overline{u}\, dx.$$

Also,

$$\int_0^t \int_{\mathbb{R}^n} L[u](s,x) \cdot \overline{u(s,x)}\, dsdx$$

$$\leqslant \frac{1}{2} \|u(t,\cdot)\|^2_{L^2(\mathbb{R}^n)} - \frac{1}{2}\|u(0,\cdot)\|^2_{L^2(\mathbb{R}^n)}$$

$$+ C \int_0^t \|u(s,\cdot)\|^2_{L^2(\mathbb{R}^n)} ds.$$

Then, from the above, we obtain the following *a priori* estimate:

$$\|u(t,\cdot)\|^2_{L^2(\mathbb{R}^n)}$$

$$\leqslant e^{Ct} \left\{ \|u(0,\cdot)\|^2_{L^2(\mathbb{R}^n)} + \int_0^t \|f(s,\cdot)\|^2_{L^2(\mathbb{R}^n)} ds \right\}.$$

For sufficiently large λ, if we can prove the existence of the solution of

$$\sum_{j=1}^n A_j(x) \frac{\partial v}{\partial x_j} + A_0(x)v - \lambda v = g$$

and estimate it; then it is possible to apply the Hille-Yosida theorem.

4. If we integrate $\mathrm{Re}\, Pu\overline{u}_t$ over $(t_1, t_2) \times \Omega$, then we get that

$$E(u; t_2) = E(u; t_1) + \int_{t_1}^{t_2} dt \int_\Omega h_0(x) \left| \frac{\partial u}{\partial t}(t,x) \right|^2 dx.$$

Now, if $E(u; t_2) = E(u; t_1)$, for $h_0 > 0$, then we obtain that $u_t(t, x) = 0$. Hence, $u_{tt}(t, x) = 0$, and then since $Pu = 0$, we have $\sum \partial_{x_j}(a_{kj} u_{x_k}) = 0$ in Ω. So, $u_{x_k}(t, x) = 0$, from which it follows that $u \equiv 0$.

Chapter 3.

1. We begin by setting $H(\tau, \xi) = \tau^2 - a_{11}\xi_1^2 - a_{22}\xi_2^2 - a_{33}\xi_3^2$. We also choose $\Phi(t, x) = t - \varphi(x)$ so that $H(\nabla\Phi) = 0$. From the results of §3.3(b), the solution of (3.2) propagates along the characteristic curve. Now, let $-\nabla\varphi(x_0) = \theta_0 = (\theta_{01}, \theta_{02}, \theta_{03})$.

Then, the characteristic curve that passes through the point x_0 is $X(t) = x_0 + ta\theta_0$, where $a\theta_0 = (a_{11}\theta_{01}, a_{22}\theta_{02}, a_{33}\theta_{03})$. This fact shows that at x_0 the wave is propagated with a speed $|a\theta_0|$ in the direction $a\theta_0$ and in a straight manner.

Now, since $a\nabla\varphi \cdot \nabla\varphi = 1$, we get that $a\theta_0 \cdot \theta_0 = 1$. The normal vector at θ_0 of the surface $\{\xi \in \mathbb{R}^3; a\xi \cdot \xi = 1\}$ is given by $a\theta_0$, and it is the direction in which the wave propagates.

Also, if we let the angle between θ_0 and $a\theta_0$ be γ, then we see that $|a\theta_0| = |\theta_0|^{-1}(\cos\gamma)^{-1}$. Since on the axis of the surface $\gamma = 0$, we must have $|a\theta_0| = |\theta_0|^{-1}$. So, if we assume that γ is not very large, the approximate speed of the wave that propagates in the direction $a\theta_0$ is inversely proportional to the size of θ_0 on the surface. That is to say, the size of θ_0 on the surface indicates the magnitude of lateness of the wave that propagates in the direction $a\theta_0$.

2. Since

$$e^{-ik(t-\varphi(x))}\operatorname{div} \boldsymbol{A}$$
$$= ik(-\operatorname{grad}\psi \cdot \boldsymbol{v}_0) + (ik)^0(-\operatorname{grad}\psi \cdot \boldsymbol{v}_1 + \operatorname{div}\boldsymbol{v}_0)$$
$$+ \ldots + (ik)^{-j+1}(-\operatorname{grad}\varphi \cdot \boldsymbol{v}_j + \operatorname{div}\boldsymbol{v}_{j-1}) + \ldots,$$

we choose $\boldsymbol{v}_j(0, x)$ in such a way that $\operatorname{grad}\psi \cdot \boldsymbol{v}_0(0, x) = 0$, $\operatorname{grad}\psi \cdot \boldsymbol{v}_1(0, x) = \operatorname{div}\boldsymbol{v}_0, \ldots$ are all satisfied. Then, we construct each component of $\boldsymbol{A}$ in such a way that it is an asymptotic solution of $\Box u = 0$.

Next, since $\operatorname{div} \boldsymbol{A}|_{t=0}$ and

$$\Box \operatorname{div} \boldsymbol{A} = \operatorname{div}\Box \boldsymbol{A} = O(k^{-N+1}),$$

we see that $\operatorname{div}\boldsymbol{A} = O(k^{-N+1})$. Now, by using (1.24) and (1.25), we see that the principal part of $\boldsymbol{B}$ is $\exp(ik - \psi(x)) \cdot (-ik)\operatorname{grad}\psi \times \boldsymbol{v}_0$. Also, the principal part of $\boldsymbol{E}$ is $\exp(ik - \psi(x))(-ik)\boldsymbol{v}_0$. Hence, the oscillations of the principal parts of both $\boldsymbol{B}$ and $\boldsymbol{E}$ are in the direction perpendicular to $\operatorname{grad}\psi$; i.e., in the direction of motion of the wave, and moreover, $\boldsymbol{B}$ and $\boldsymbol{E}$ are mutually orthogonal.

3. Consider the case of the wave that is the asymptotic solution of (3.187) with $\lambda = \lambda_1$ and is transversely incident at the boundary. That is, we assume that $-\Delta\varphi \cdot \nu(x) \geqslant \alpha_0 > 0$. Further, we assume that the reflected wave is formed by means

of

$$\boldsymbol{u}^+ = e^{ik(\lambda_1 t - \varphi^+(x))} \sum_{l=0}^{N} (ik)^{-l} \boldsymbol{v}_l^+(t, x),$$

$$\boldsymbol{w}^- = e^{ik(\lambda_1 t - \psi(x))} \sum_{l=0}^{N} (ik)^{-l} \boldsymbol{r}_l(t, x),$$

where the phase function is chosen so that the following are satisfied:

$$\begin{cases} |\nabla \varphi^+| = 1, \quad |\nabla \psi|^2 = (\alpha + \beta)/\alpha \quad \text{in } \Omega, \\ \varphi^+(x) = \psi(x) = \varphi(x) \quad \text{on } \Gamma. \end{cases}$$

If such a phase function is found, then it suffices to determine the value of the amplitude function on Γ so that the boundary condition is satisfied.

Now, look at the case of $l = 0$. Since $\boldsymbol{v}_0$ is of the form $p_0 \nabla \varphi$, it is sufficient to find a real-valued function p_0 and a vector-valued function $\boldsymbol{r}_0$, which is orthogonal to $\nabla \psi$, such that they satisfy

$$p_0 \nabla \varphi = p_0^+ \nabla \varphi^+ + \boldsymbol{r}_0 \quad \text{on } \Gamma.$$

By setting $\boldsymbol{r}_0 = q\boldsymbol{h} + r\boldsymbol{j}$, we need to check whether (p^+, q, r) is determined.

4. At time $t = 0$ we assume that the amplitude function v is of the form $v(0, x; k) = m(|x|)p(\omega)$, where $\omega = x/|x|$. Further, suppose that supp $m = \{l; 1 \leqslant l \leqslant 2\}$. Then, from the transport equation, we obtain that $v_0(t, x) = (|x|+t)/|x| \cdot m(|x|+t)p(\omega)$. By letting $v_1(0, x) = 0$, we can determine v_1. If P is not a harmonic function on the sphere, then there exists a $c_0 > 0$ such that

$$\sup_{|x|=r} |\Box v_0(t, x)| \geqslant c_0 |m(t + r)| r^{-3} + O(r^{-2}).$$

Therefore,

$$\sup_{|x|=r} |v_1(t, x)| \geqslant \tilde{c}_0 |m(t + r)| r^{-2} + O(r^{-1}).$$

By continuing this approach, we shall eventually obtain that the following holds true: $\sup_{|x|=r} |v_j(t, x)| \geqslant c_j |m(t+|x|)| r^{-j-1} + O(r^{-j}).$

Hence, for some $c_N > 0$, we see that

$$\sup_{|x|=r} |\Box u(t, x; k)| \geqslant c_N |m(t + |x|)| k^{-N} r^{-N}.$$

Now, if we consider the above estimate and the conditions on the support of m and, in addition, we also consider the upper bound estimate of the above, then in the region $t < 1 - k^{-(1-\varepsilon)}$ ($\varepsilon > 0$) as $k \to \infty$ we see that $\Box u$ becomes small. However, for $t \geqslant 1 - k^{-1}$, the estimate is difficult.

Clearly, at $t = 1$, the parts that are in $|x| = 1$ at $t = 0$ are clustered at $x = 0$, so we can say $x = 0$ is the focus. The asymptotic solution of the form in (3.96) is also valid in the region that is not at the focus, but close to the focus (the actual distance depends on the frequency k) it is not valid. So far, there is no general theory for the construction of an asymptotic solution in the domain that contains the focus.

Chapter 4.

1. Let us suppose that on the ray that emerges from a point of $\mathcal{R}$ there is a point X that passes through FA and a point Y that also lies on the curve ABC. The segment XY is to the right of the segment FY. Hence, the reflected ray at Y is to the left of $F'Y$, and never passes through the segment FF'.

2. By integrating $Pu \cdot u_t$ over $(0, t) \times \mathbb{R}^n$, we obtain that

$$\int_0^t \alpha \|u(s, \cdot)\|_{L^2(\mathbb{R}^n)}^2 ds \leqslant E(0; u, \infty).$$

Then, by choosing a suitable λ and by integrating $Pu \cdot (tu_t + \lambda u)$, we see that

$$\int_0^t sE(s; u, \infty) ds \leqslant CE(0; u, \infty).$$

Index

Selected Titles in This Series

(Continued from the front of this publication)

For a complete list of titles in this series, visit the
AMS Bookstore at **www.ams.org/bookstore/**.